Intermediate Electromagnetic Theory

Douglas A. Barlow

© 2023, Douglas A. Barlow, All rights reserved.

ISBN: 978-1-304-89465-6
Lulu Publishing, Research Triangle, NC

© 2023, Douglas A. Barlow, All rights reserved.

Contents

Contents		**1**
1 Vector Analysis		**5**
1.1	Vector Notation	5
1.2	The Vector Inner Product	8
1.3	The Vector Cross Product	12
1.4	Vector Triple Products	19
1.5	Vector Fields	22
1.6	Flux of a Vector Field	34
1.7	Important Integral Theorems	37
1.8	Cylindrical Coordinates	43
1.9	Spherical Coordinates	48
2 Electrostatics		**55**
2.1	Coulomb's Law	55
2.2	The Electric Field	61
2.3	Gauss's Law	69
2.4	The Electrostatic Field	77

3 Voltage and Capacitance 83
- 3.1 Potential Energy of Static Charges 83
- 3.2 Electric Potential Difference 90
- 3.3 Capacitance 98
- 3.4 Energy and Charge Density 101
- 3.5 Energy in Electric Fields 104

4 Electric Fields in Matter 111
- 4.1 Dielectrics 111
- 4.2 Bound Charges 115
- 4.3 Linear Dielectrics 120
- 4.4 The Displacement Field 125
- 4.5 Electrostatic Boundary Conditions 130
- 4.6 Energy in Dielectrics 133
- 4.7 The Clausius-Mossotti Equation 136

5 Magnetostatics 143
- 5.1 Current 143
- 5.2 Drude Theory of Conductivity 145
- 5.3 Resistance 150
- 5.4 The Biot-Savart Law 159
- 5.5 Magnetic Forces 164
- 5.6 The Velocity Selector 172
- 5.7 The Magnetic Moment 178
- 5.8 Ampère's Law 184
- 5.9 The Solenoid 189

6 Magnetic Fields in Matter 197
- 6.1 Magnetization 197
- 6.2 Bound Currents 200
- 6.3 Ampère's Law in Materials 202
- 6.4 Linear Media 205
- 6.5 Magnets 210

CONTENTS

7 Faraday's Law — **217**
- 7.1 Magnetic Flux 217
- 7.2 Electromagnetic Induction 219
- 7.3 Induction in a Moving Conductor 223
- 7.4 Self Inductance 227
- 7.5 Lenz's Law 230
- 7.6 Energy in a Magnetic Field 235
- 7.7 Mutual Inductance 236
- 7.8 Magnetic Circuits 243

8 Maxwell's Equations — **253**
- 8.1 Displacement Current 253
- 8.2 The Wave Equation 259
- 8.3 Energy in E & M Waves 264
- 8.4 Maxwell's Equations in Materials 266
- 8.5 Maxwell's Equations and Conductors . . . 270
- 8.6 Electric Dipole Radiation 274

Appendices — **281**
- A Integral Formulas 281
- B Trigonometric Identities 284
- C Useful Relations 285
- D Relations Involving the Del Operator . . . 286
- E Del in Curvlinear Coordinates 287
- F Physical Constants 288
- G INDEX 289

Chapter 1

Vector Analysis

1.1 Vector Notation

During this course in electromagnetics we will encounter many vector fields. It is then essential that the concept of a vector be reviewed here and that operations with vectors, which will be useful later in this text and that may be new to the reader, be introduced.

A *vector* is a mathematical entity that holds a magnitude and a sense of direction. A quantity fully described by a magnitude alone, without any sense of direction, we call a *scalar*.

In two dimensions, it is traditional to denote some vector $\vec{A}$ as

$$\vec{A} = A_x \, \hat{x} + A_y \, \hat{y} \, . \qquad (1.1)$$

Here the scalar quantities A_x and A_y are referred to as the horizontal and vertical components of $\vec{A}$ respectively. $\hat{x}$ and $\hat{y}$ are called unit vectors. All unit vectors are unitless and have magnitude one. A unit vector $\hat{A}$ (read, A hat) in the direction of the vector $\vec{A}$ is given by

$$\hat{A} = \frac{\vec{A}}{|\vec{A}|}, \qquad (1.2)$$

where $|\vec{A}|$ denotes the *magnitude* or length of $\vec{A}$ and is given by

$$|\vec{A}| = \sqrt{A_x{}^2 + A_y{}^2}. \qquad (1.3)$$

$\hat{x}$ and $\hat{y}$ are a special pair of unit vectors that form what we call a *complete orthogonal basis* for vectors in two dimensions. It is a straight forward thing to define another unit vector $\hat{z}$ and then have such a basis for three dimensional vector space. The fact that the basis is complete, implies that any vector in three dimensions can be written as a linear combination of these three vectors. That is

$$\vec{A} = A_x\,\hat{x} + A_y\,\hat{y} + A_z\,\hat{z}, \qquad (1.4)$$

where the components A_x, A_y and A_z could be constants or functions of time and/or space, thus making $\vec{A}$ a *vector valued function*. The orientation of the basis vectors for three dimensional Cartesian vector space is shown in Figure 1.1.

There are similar sets of basis vectors that do the same job for other coordinates systems such as for cylindrical or spherical. Coordinate systems of this type are referred to as *curvilinear*. Unfortunately, unlike in the Cartesian system, basis vectors in these systems vary with the direction of the vector. One can see this is the case by considering a vector in spherical coordinates, along with its corresponding basis set, as shown in Figure 1.2. This fact makes analysis with vectors in these curvilinear systems

1.1. VECTOR NOTATION

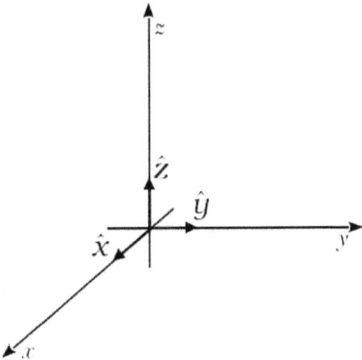

Figure 1.1: Three dimensional Cartesian coordinate system with the three basis vectors shown.

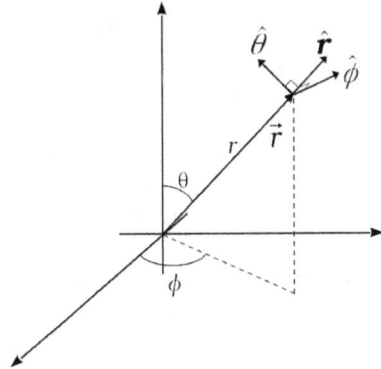

Figure 1.2: Three dimensional spherical coordinate system showing a vector $\vec{r}$ along with its three basis vectors.

cumbersome to say the least. In this text we will avoid vectors outside of the Cartesian system for the most part. We will however use the cylindrical and spherical systems frequently when dealing with certain scalar functions.

1.2 The Vector Inner Product

In the section we will discuss a binary multiplication operation that is defined for vectors, the *inner* or *dot product*. In doing so, some of the concepts mention in the previous section will be further clarified.

Consider the vectors $\vec{A}$ and $\vec{B}$ in n dimensional vector space. The inner product of $\vec{A}$ and $\vec{B}$, denoted as $\vec{A} \cdot \vec{B}$ is given by

$$\vec{A} \cdot \vec{B} = A_1 B_1 + A_2 B_2 + A_3 B_3 + \cdots + A_n B_n \ . \quad (1.5)$$

In this text we will restrict our use of vectors to situations where $1 \leq n \leq 3$. The inner product of two vectors yields a scalar.

The vector inner product is commutative:

$$\vec{A} \cdot \vec{B} = \vec{B} \cdot \vec{A} \ . \quad (1.6)$$

The vector inner product is distributive:

$$\vec{A} \cdot (\vec{B} + \vec{C}) = \vec{A} \cdot \vec{B} + \vec{A} \cdot \vec{C} \ . \quad (1.7)$$

The inner product of a vector with itself is called the *norm* of the vector. Therefore, the norm of vector $\vec{A}$ in three dimensional space is given by

$$\vec{A} \cdot \vec{A} = A_x^2 + A_y^2 + A_z^2 \ . \quad (1.8)$$

It can be seen, from Eq. (1.3) in the previous section, that the magnitude of a vector is simply the square root of the norm.

The following relation, referred to as the *Schwarz inequality*, holds for the inner product of two vectors $\vec{A}$ and $\vec{B}$:

$$(\vec{A} \cdot \vec{B})^2 \leq (\vec{A} \cdot \vec{A})(\vec{B} \cdot \vec{B}) \ . \quad (1.9)$$

1.2. THE VECTOR INNER PRODUCT

If any two vectors $\vec{A}$ and $\vec{B}$ are orthogonal (perpendicular) then
$$\vec{A} \cdot \vec{B} = 0 . \tag{1.10}$$
Therefore, for our three dimensional unit basis vectors we must have that
$$\hat{x} \cdot \hat{x} = \hat{y} \cdot \hat{y} = \hat{z} \cdot \hat{z} = 1 , \tag{1.11}$$
and, since they are orthogonal,
$$\hat{x} \cdot \hat{y} = \hat{y} \cdot \hat{z} = \hat{x} \cdot \hat{z} = 0 . \tag{1.12}$$
Given a complete orthogonal basis $\{\hat{e}_n\}$ in n dimensional vector space the nth component of some vector $\vec{A}$ can then be given by
$$A_n = \vec{A} \cdot \hat{e}_n . \tag{1.13}$$

The inner product of two vectors has a geometric description. Consider the two vectors $\vec{A}$ and $\vec{B}$ as shown in Figure 1.3. By Eq. (1.5) we must have that

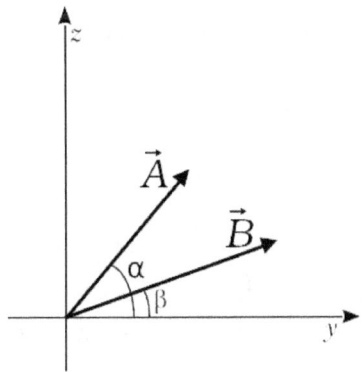

Figure 1.3: Two vectors in two dimensions.

$$\vec{A} \cdot \vec{B} = A_x B_x + A_y B_y \ . \qquad (1.14)$$

From Figure 1.3, it can be seen that the horizontal and vertical components of $\vec{A}$ and $\vec{B}$ can be written in terms of trigonometric functions as

$$\vec{A} \cdot \vec{B} = (A\cos\alpha)(B\cos\beta) + (A\sin\alpha)(B\sin\beta) \ , \qquad (1.15)$$

where the magnitudes of $\vec{A}$ and $\vec{B}$ are given by A and B respectively. Factoring this and using a trigonometric identity leads to

$$\vec{A} \cdot \vec{B} = AB \left[\frac{1}{2}\cos(\alpha - \beta) + \frac{1}{2}\cos(\alpha + \beta) + \frac{1}{2}\cos(\alpha - \beta) - \frac{1}{2}\cos(\alpha + \beta) \right] \ . \qquad (1.16)$$

This reduces to

$$\vec{A} \cdot \vec{B} = AB\cos(\alpha - \beta) \ . \qquad (1.17)$$

Here, $\alpha - \beta$ is the angle between the vectors $\vec{A}$ and $\vec{B}$. Letting $\theta = \alpha - \beta$ we can then write that

$$\vec{A} \cdot \vec{B} = AB\cos\theta \ , \qquad (1.18)$$

so if the angle between two vectors is known, along with their magnitudes, then the inner product can be computed using Eq. (1.18). This result was derived for the two-dimensional case but can be shown to be correct for any two vectors in three dimensions as well.

1.2. THE VECTOR INNER PRODUCT

Example 1.1

The potential energy U for a magnetic moment vector, $\vec{\mu}$, at a point in space where the magnetic field is given by the vector $\vec{B}$, is given by

$$U = -\vec{\mu} \cdot \vec{B} .$$

Suppose $\vec{\mu} = 2.0\ \hat{x}$ Am^2 and $\vec{B} = a\ \hat{x} + b\ \hat{y}$ T. Here, a and b are constants. If $U = 0.1$ J, what restrictions are placed on a and b?

Using Eq. (1.5) leads to

$$0.1 = -\vec{\mu} \cdot \vec{B} = -2a - 0 \cdot b ,$$

so that $a = -0.1/2 = -0.05$ T and b can be any real number.

PROBLEMS

1.1 Verify the Schwarz inequality (Eq. (1.9)) for two vectors in two dimensions.

1.2 Verify Eq. (1.6) for vectors in two dimensions.

1.3 Verify Eq. (1.7) for vectors in two dimensions.

1.4 In mechanics, the work, W, done by some force $\vec{F}$ on moving an object through a straight line displacement $\vec{d}$ is given by

$$W = \vec{F} \cdot \vec{d} \, .$$

Suppose $\vec{d} = 2.0\,\hat{x} - 5.0\,\hat{y} - 4.0\,\hat{z}$ m. Compute the work done by the following forces during this displacement.

a) $\vec{F} = 2.2\,\hat{x}$ N b) $\vec{F} = -4.0\,\hat{x} - 2.0\,\hat{z}$ N
c) $\vec{F} = -6.0\,\hat{x} + 2.0\,\hat{y} + 5.0\,\hat{z}$ N

What force, $\vec{F}$, yields $W = 0$?

1.3 The Vector Cross Product

Due to the development of definitions like torque and angular momentum, it has been useful to define another type of vector multiplication which yields a vector as its product. As with the inner product, we will only be concerned with vectors spaces of dimension n where $1 \leq n \leq 3$. We refer to this operation as the *vector* or *cross product*.

Consider two vectors $\vec{A}$ and $\vec{B}$. We denote the cross product of these two vectors as

$$\vec{C} = \vec{A} \times \vec{B} \, .$$

Rather than associate the magnitude of this product with the cosine of the angle, θ, between the two vectors, as with the inner product in Eq. (1.18), the magnitude of the product, C, is given by

$$C = AB \sin \theta \, . \tag{1.19}$$

Further, the product vector $\vec{C}$ is defined to be normal to the plane described by $\vec{A}$ and $\vec{B}$. Therefore, for the three

1.3. THE VECTOR CROSS PRODUCT

dimensional Cartesian basis vectors it must be that

$$\hat{x} \times \hat{y} = \hat{z}, \quad \hat{y} \times \hat{z} = \hat{x}, \quad \text{and} \quad \hat{z} \times \hat{x} = \hat{y} . \quad (1.20)$$

Further, due to Eq. (1.19), we must have that,

$$\hat{x} \times \hat{x} = 0, \quad \hat{y} \times \hat{y} = 0, \quad \text{and} \quad \hat{z} \times \hat{z} = 0 . \quad (1.21)$$

The cross product is distributive,

$$\vec{A} \times (\vec{B} + \vec{C}) = \vec{A} \times \vec{B} + \vec{A} \times \vec{C} . \quad (1.22)$$

However, the cross product is not commutative. In fact it is *anticommutative*:

$$\vec{A} \times \vec{B} = -(\vec{B} \times \vec{A}) . \quad (1.23)$$

This then implies that

$$\hat{y} \times \hat{x} = -\hat{z}, \quad \hat{z} \times \hat{y} = -\hat{x}, \quad \text{and} \quad \hat{x} \times \hat{z} = -\hat{y} . \quad (1.24)$$

If $\vec{A} \times \vec{B} = \vec{C}$ and c is a constant, then

$$c\vec{A} \times \vec{B} = c\vec{C} . \quad (1.25)$$

The above rules lead to a method for computing the cross product. To see how this is so, consider the cross product $\vec{C} = \vec{A} \times \vec{B}$, where the vectors are given in component form:

$$(C_x \hat{x} + C_y \hat{y} + C_z \hat{z}) = \quad (1.26)$$
$$(A_x \hat{x} + A_y \hat{y} + A_z \hat{z}) \times (B_x \hat{x} + B_y \hat{y} + B_z \hat{z}) .$$

Using Eqs. (1.20) through (1.22) and Eq. (1.24) in the above leads to

$$C_x = A_y B_z - A_z B_y ,$$
$$C_y = A_x B_z - A_z B_x ,$$
$$C_z = A_x B_y - A_y B_x . \quad (1.27)$$

This result means that we can write the cross product of two vectors as the determinant of a particular matrix:

$$\vec{A} \times \vec{B} = \begin{vmatrix} \hat{x} & \hat{y} & \hat{z} \\ A_x & A_y & A_z \\ B_x & B_y & B_z \end{vmatrix}. \qquad (1.28)$$

Notice that the above formula involves all of the unit basis vectors for three dimensional space. Even if $\vec{A}$ and $\vec{B}$ are two dimensional in $\hat{x}$ and $\hat{y}$, the geometrical nature of the cross product requires that the $\hat{z}$ basis vector be involved.

Before we compute a cross product using Eq. (1.28), it would be helpful to review the concept of a square matrix and its determinant. A matrix is a collection of elements. In a mathematical sense, these elements might be numbers, functions or even vectors. This collection of elements are arrayed in a table-like structure for convenience and for purposes of analysis. For example, consider the four elements consisting of the integers from 1 to 4. These elements are listed in a matrix below which we will call A:

$$A = \begin{bmatrix} 1 & 2 \\ 3 & 4 \end{bmatrix}. \qquad (1.29)$$

In this matrix there are *rows* and *columns*. Rows are along the horizontal while columns the vertical. Here the elements 1 and 2 are in the first row of matrix A while 2 and 4 are in the second column. This particular matrix has 2 rows and 2 columns. We would call A a 2 by 2 (2×2) matrix. If in a matrix the number of rows equals the number of columns we say that it is a *square matrix*.

It is known that the human mind deals in a favorable way with elements tabled in a matrix. So much so, that

1.3. THE VECTOR CROSS PRODUCT

there are many different mathematical operations defined for matrices and many resulting practical applications.

Here, we are primarily concerned with one operation. That is, finding the *determinant* of a matrix. The determinant is a unique value associated with any square matrix. To denote that we want the determinant of matrix A we would write $\det(A)$ or

$$\det(A) = \begin{vmatrix} 1 & 2 \\ 3 & 4 \end{vmatrix}.$$

Note how the brackets used in Eq. (1.29) are replaced by straight vertical lines to indicate that we want the determinant of A. For a 2×2 matrix, there is a simple formula that yields the determinant:

$$\det(A) = \begin{vmatrix} a & b \\ c & d \end{vmatrix} = ad - cb . \quad (1.30)$$

This is great, but we want to be able to find the determinant of any square matrix. Therefore, a method is introduced below for doing just that.

The determinant of any square matrix can be computed by using a method known as *expansion by cofactors* or the *Laplace expansion*. To fully understand this method, we first must consider a few additional aspects of matrices. Consider the 3×3 matrix shown below:

$$A = \begin{bmatrix} 1 & 2 & 3 \\ 4 & 5 & 6 \\ 3 & 2 & 1 \end{bmatrix} . \quad (1.31)$$

Suppose we want to refer to a particular element in the matrix A given by expression (1.31). This can be done with the notation A_{ij}. This tells us that we are referring to an element in matrix A on row i and column j. For example, $A_{11} = 1$ and $A_{21} = 4$. Now, the matrix A contains several *sub-matrices*. For instance, if we disregard the elements in the row and column of A_{11} we are left with the following 2×2 sub-matrix:

$$\begin{bmatrix} 5 & 6 \\ 2 & 1 \end{bmatrix}.$$

If we find the determinant of this sub-matrix we have what is called the minor, M_{11}, of A_{11} that is,

$$M_{11} = \begin{vmatrix} 5 & 6 \\ 2 & 1 \end{vmatrix} = (5)(1) - (2)(6) = -7 \ .$$

Furthermore, if the minor of an element is multiplied by $(-1)^{i+j}$ we have what is then known as the *cofactor* of the element. So, the cofactor C_{11} of the element A_{11} would be

$$C_{11} = (-1)^{1+1}(-7) = -7 \ .$$

We are now able to discuss a procedure for finding the determinant of any square matrix. The determinant of a square matrix is found by summing the product of an element with its cofactor along any one row or column. As a formula, for any $N \times N$ matrix A, this statement is

$$\det(A) = \sum_{i \text{ or } j=1}^{N} (-1)^{i+j} M_{ij} \ . \qquad (1.32)$$

1.3. THE VECTOR CROSS PRODUCT

Example 1.2

Use formula (1.32) to find the determinant of the following matrix:

$$A = \begin{bmatrix} 2 & 6 & 2 \\ 3 & 2 & 0 \\ 2 & 2 & 0 \end{bmatrix}.$$

Here there are two elements of zero in the third column so then our sum along the third column will have only one non-zero term. So,

$$\begin{vmatrix} 2 & 6 & 2 \\ 3 & 2 & 0 \\ 2 & 2 & 0 \end{vmatrix} = 2(-1)^{1+3} \begin{vmatrix} 3 & 2 \\ 2 & 2 \end{vmatrix} + 0 + 0 =$$

$$2\left[(3)(2) - (2)(2)\right] = 4 \ .$$

The determinant for the above matrix is 4.

You may have noticed by now that the factor of $(-1)^{i+j}$ leads to either a 1 or -1 in the cofactor. Often, when dealing with a 3×3 matrix it is convenient to remember the following pattern:

$$\begin{bmatrix} + & - & + \\ - & + & - \\ + & - & + \end{bmatrix}.$$

Now, if we are dealing with a cofactor in a plus position we know that $(-1)^{i+j}$ leads to +1 while in a negative position $(-1)^{i+j}$ gives -1.

Example 1.3

Consider the following two vectors:

$$\vec{v} = 2\,\hat{x} + 4\,\hat{y}$$
$$\vec{u} = -3\,\hat{x} + 2\,\hat{y}\ .$$

Find $\vec{v} \times \vec{u}$.

We set up the required determinant using formula (1.28):

$$\vec{v} \times \vec{u} = \begin{vmatrix} \hat{x} & \hat{y} & \hat{z} \\ 2 & 4 & 0 \\ -3 & 2 & 0 \end{vmatrix}.$$

With two zeros in the third column we expand by cofactors along the third column:

$$\begin{vmatrix} \hat{x} & \hat{y} & \hat{z} \\ 2 & 4 & 0 \\ -3 & 2 & 0 \end{vmatrix} = \hat{z}[2(2) - (-3)(4)] - 0 + 0 = 16\,\hat{z}\ .$$

Notice how the resulting vector is in the $\hat{z}$ direction. That is, the product is perpendicular or normal to the $\hat{x}$ and $\hat{y}$ (xy) plane.

PROBLEMS

1.5 Verify Eq. (1.22) for vectors in two dimensions.

1.6 Verify Eq. (1.23) for vectors in two dimensions.

1.7 Verify Eq. (1.25) for vectors in two dimensions.

1.8 Derive Eqs. (1.27) from Eq. (1.26).

1.9 The force, $\vec{F}$, on a particle of charge q can be given by $\vec{F} = q\vec{v} \times \vec{B}$. If $\vec{v} = v_x\,\hat{x} + v_y\,\hat{y}$ and $\vec{B} = B_o\,\hat{z}$ find $\vec{F}$. (Assume all SI units are used.)

1.4 Vector Triple Products

It is worth discussing some special cases involving the inner and cross product that involve three factors. For example,
$$\vec{A} \cdot (\vec{B} \times \vec{C}) \,, \qquad (1.33)$$
is sometimes referred to as the *scalar triple product*. Now, since $\vec{B}$ and $\vec{C}$ define a plane in space, and assuming $\vec{A}$ does not lie within this plane, then $\vec{A}$, $\vec{B}$ and $\vec{C}$ give the sides of a three dimensional shape called a *parallelepiped* as shown in Figure 1.4. It can be shown that $|\vec{A} \cdot (\vec{B} \times \vec{C})|$ gives the volume of the parallelepiped described by $\vec{A}$, $\vec{B}$ and $\vec{C}$.

The scalar triple product is commutative if the " alphabetical" order of the vectors is maintained. That is,
$$\vec{A} \cdot (\vec{B} \times \vec{C}) = \vec{B} \cdot (\vec{C} \times \vec{A}) = \vec{C} \cdot (\vec{A} \times \vec{B}) \,. \qquad (1.34)$$

From Eq. (1.23) we know that

$$\vec{A} \cdot (\vec{C} \times \vec{B}) = \vec{B} \cdot (\vec{A} \times \vec{C}) = \vec{C} \cdot (\vec{B} \times \vec{A}) , \qquad (1.35)$$

would yield $-\vec{A} \cdot (\vec{B} \times \vec{C})$. Also, the order of the inner and cross product operations can be interchanged:

$$\vec{A} \cdot (\vec{B} \times \vec{C}) = (\vec{A} \times \vec{B}) \cdot \vec{C} . \qquad (1.36)$$

It can be shown that

$$\vec{A} \cdot (\vec{B} \times \vec{C}) = \begin{vmatrix} A_x & A_y & A_z \\ B_x & B_y & B_z \\ C_x & C_y & C_z \end{vmatrix} . \qquad (1.37)$$

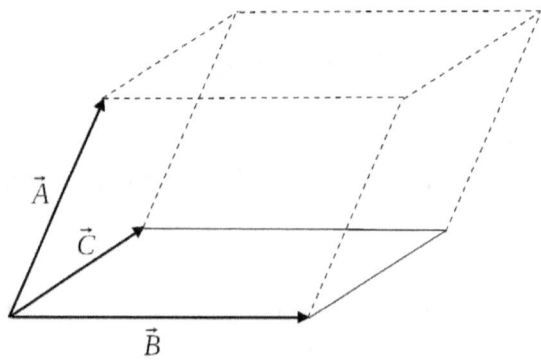

Figure 1.4: A parallelepiped described by the three vectors $\vec{A}$, $\vec{B}$ and $\vec{C}$.

Another important triple product is the *vector triple product*: $\vec{A} \times (\vec{B} \times \vec{C})$. This triple product can be simplified by using the world famous BAC-CAB rule:

$$\vec{A} \times (\vec{B} \times \vec{C}) = \vec{B}(\vec{A} \cdot \vec{C}) - \vec{C}(\vec{A} \cdot \vec{B}) . \qquad (1.38)$$

1.4. VECTOR TRIPLE PRODUCTS

The triple products discussed in this section can be used to derive the following useful vectors product identities:

$$(\vec{A} \times \vec{B}) \cdot (\vec{C} \times \vec{D}) = (\vec{A} \cdot \vec{C})(\vec{B} \cdot \vec{D}) - (\vec{A} \cdot \vec{D})(\vec{B} \cdot \vec{C}). \quad (1.39)$$

$$\vec{A} \times (\vec{B} \times (\vec{C} \times \vec{D})) = \quad (1.40)$$
$$\vec{B}(\vec{A} \cdot (\vec{C} \times \vec{D})) - (\vec{A} \cdot \vec{B})(\vec{C} \times \vec{D}).$$

$$(\vec{A} \times (\vec{B} \times \vec{C})) + (\vec{B} \times (\vec{C} \times \vec{A})) + (\vec{C} \times (\vec{A} \times \vec{B})) = 0. \quad (1.41)$$

A larger listing of vector identities can be found in an Appendix.

PROBLEMS

1.10 Verify Eq. (1.36) for vectors in two dimensions.

1.11 Verify Eq. (1.37).

1.12 Verify the BAC-CAB rule (Eq. (1.38)) for vectors in two dimensions.

1.13 Find the volume of the parallelpiped describe by the vectors $\vec{A} = 2.0\,\hat{x} - 4.0\,\hat{y}$ m, $\vec{B} = 2.0\,\hat{x} - 6.0\,\hat{z}$ m and $\vec{C} = 3.0\,\hat{x} - 4.0\,\hat{y} + 5.0\,\hat{z}$ m.

1.5 Vector Fields

Though vectors can be used to describe the velocity, acceleration or momentum for a lone object, it is often useful to use vectors to describe collective phenomena involving many particles. A good example of such a situation would be the flow of particles in a liquid. Rather than list the velocity vectors for each particle one simply refers to a single vector valued expression that can then be used to assign a velocity to any particle in the area of interest.

For example, if the liquid is flowing to the right with all particles having the same speed of 2.3 m/s, a velocity vector field could be written as $\vec{v} = 2.3\,\hat{x}$ m/s. Here, its as if every point in space can be assigned the vector $\vec{v}$. Of course usually, we are only interested in the particles in our particular system of interest and even then, we often only denote a few of them so as to visually convey the essence of the flow field. We see examples of $\vec{v}$ denoted for selected particles in a liquid flowing to the right through a section of pipe in Figure 1.5.

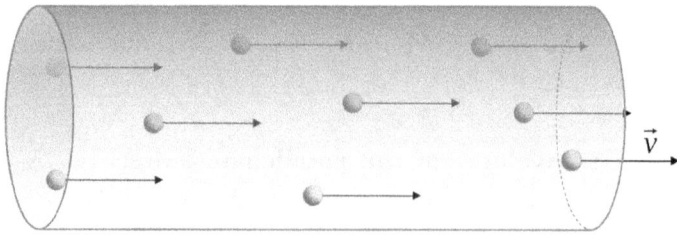

Figure 1.5: Liquid flow in a pipe described a vector field, $\vec{v}$, of particles moving to the right at speed $|\vec{v}|$.

In the above case, the vectors were defined over all space even though we were only interested in those within

1.5. VECTOR FIELDS

a certain region. This is usually the situation. Also, it is important to note, that this vector field was composed of vectors all having the same magnitude and direction. We will refer to such vector fields as *uniform*.

It could be that the vectors in the field vary with position and time. For now, we will focus on those fields that may, or may not, vary with position but are constant with respect to time. We call such vector fields, *static fields*. For example, consider the vector field

$$\vec{A}(y) = A_o e^{-k|y|} \hat{x} , \qquad (1.42)$$

where both A_o and k are positive constants. This vector field contains vectors that are all directed to the right, as was the case in Figure 1.4, however here the magnitude of the vectors in the field decreases as one moves away from the x axis. That is, for every point (x_i, y_i) in two dimensional space, there is a vector at that point given by $\vec{A}(y_i) = A_o e^{-k|y_i|} \hat{x}$.

The Vector Gradient

When one has a vector field where the field vector components vary over the area of interest it is useful to quantify this change with respect to a given change in position in the field. That is, we might be interested in $d\vec{A}/dy$. Using Eq. (1.42) as an example this gives

$$\frac{d\vec{A}}{dy} = -kA_o e^{-k|y|} \hat{x} . \qquad (1.43)$$

Not surprisingly, the derivative of a vector valued function leads to another vector valued function. However, it can be useful to have the change with respect to position as a scalar function. One prescription for obtaining such a thing would be the following inner product

$$\frac{d}{dy}\hat{x} \cdot A_o e^{-k|y|}\hat{x} \,.\qquad(1.44)$$

Here our first product is a vector whose lone component is an operator.

As you might imagine, many different vectors operators of this sort could be written. One that is very useful is the following:

$$\vec{\nabla} = \frac{\partial}{\partial x}\hat{x} + \frac{\partial}{\partial y}\hat{y} + \frac{\partial}{\partial z}\hat{z}\,.\qquad(1.45)$$

Here the symbol $\vec{\nabla}$ is called *del*. In fact, we refer to Eq. (1.45) as the del operator in Cartesian coordinates. We can perform binary operations with $\vec{\nabla}$, and any other vector, following the rules outline in this chapter. However, it is useful first to consider simply letting $\vec{\nabla}$ operate on some scalar function of position. For example, let $f = f(x, y, z)$ then the operation $\vec{\nabla} f$ yields a vector. This operation on f is called the *gradient* of f. It has a geometric interpretation. $\vec{\nabla} f$ points in the direction of maximum increase of f.

Example 1.4

Compute the gradient of $f(x, y, z) = 5x + 6y^2 + 2\sin(kz)$ where k is some constant.

We use Eq. (1.45) to write

1.5. VECTOR FIELDS

$$\vec{\nabla}f = (\tfrac{\partial}{\partial x}\hat{x} + \tfrac{\partial}{\partial y}\hat{y} + \tfrac{\partial}{\partial z}\hat{z})(5x + 6y^2 + 2\sin(kz)) \ .$$

Distribute through to get,

$$\vec{\nabla}f = \tfrac{\partial}{\partial x}(5x + 6y^2 + 2\sin(kz))\,\hat{x} +$$
$$\tfrac{\partial}{\partial y}(5x + 6y^2 + 2\sin(kz))\,\hat{y} + \tfrac{\partial}{\partial z}(5x + 6y^2 + 2\sin(kz))\,\hat{z} \ .$$

Performing the indicated differentiation leads to

$$\vec{\nabla}f = 5\,\hat{x} + 12y\,\hat{y} + 2k\cos(kz)\,\hat{z} \ .$$

For any selected (y, z), this vector points in the direction of maximum increase of f from that position.

PROBLEMS

1.14 Compute the gradient of the following functions.

a) $f(x,y) = \sqrt{x^2 + y^3}$.
b) $g(x,y,z) = A_o y e^{-kx^2} \sin(kz)$, where A and k are constants.
c) $g(x,y,z) = xy^{1/3} + 5z^3 + 4x(y+1)$.
d) $f(x,y) = 5y^2 \sinh(kx^2) + 6\cosh(x+y)$

1.15 The topography of a hill can be describe by the following equation:

$$h(x,y) = 10(2xy - 3x^2 - 4y^2 - 18x + 28y + 12) \ .$$

Here h is in feet but x and y are in miles. If one is standing at the location (1.2, 2.3) what is the steepest direction from this point?

The Divergence

In the previous section, we learned about the operation of del on a scalar function. In view of topics covered earlier in this chapter, we would like to now consider the inner, or dot, product of del with another vector valued function, say $\vec{A}$. That is, in three dimensions

$$\vec{\nabla} \cdot \vec{A} = \frac{\partial A_x}{\partial x} + \frac{\partial A_y}{\partial y} + \frac{\partial A_z}{\partial z} . \qquad (1.46)$$

We refer to this operation as the divergence of $\vec{A}$. What does Eq. (1.46) tell us about the vector field $\vec{A}$ and how is this knowledge useful? First, it can be seen from the right side of Eq. (1.46) that if all of the components of $\vec{A}$ are constants or somehow the sum on the right side of Eq. (1.46) equals zero then $\vec{\nabla} \cdot \vec{A} = 0$. Therefore, it must be that the vector field has a non-zero divergence when there is a net change in the vector components with respect to the coordinate of their basis vector.

To arrive at a physical description for the divergence let's consider the flow of particles in a fluid. It is obvious, that for the velocity vector field given in Figure 1.5, and discussed previously, that the divergence of this vector field is zero. However, it could be that the velocity field varies with position, that is $\vec{v} = \vec{v}(x, y, z)$. Now, if the mass density of the fluid is given by ρ, where $\rho = \rho(x, y, z, t)$, where t represents time, then we can define something we call the particle flux $\vec{J}_v$:

$$\vec{J}_v = \rho \vec{v} . \qquad (1.47)$$

$\vec{J}_v$ gives the mass flowing through unit area per unit time. If $\vec{\nabla} \cdot \vec{J}_v \neq 0$ then it must be that the fluid density is

1.5. VECTOR FIELDS

changing over time so that

$$\vec{\nabla} \cdot \vec{J}_v = -\frac{\partial \rho}{\partial t}. \tag{1.48}$$

Eq. (1.48) is an example of a *continuity equation*. It is a statement of the general principle of conservation of mass. Note that if $\vec{\nabla} \cdot \vec{J}_v > 0$ then there is a net flow of mass out of some region since it must be that $\partial \rho / \partial t < 0$ in this case. Likewise, if $\vec{\nabla} \cdot \vec{J}_v < 0$ then mass is flowing into the region of interest and $\partial \rho / \partial t > 0$. When there is a net flow of mass out of a region we say there is a *source*, or *sources*, present. When there is a net flow of mass into a region we say there is a *sink*, or *sinks*, present.

Now there could be both sources and sinks present in the region of interest. When the sinks and sources do not exactly cancel one another, that is if overall the vector flux leaving sources is not exactly equal to the flux going into sinks, then the divergence of the vector field will be non-zero. However, in the case where the divergence of a vector field is zero, one can think of this situation as being where sources and sinks are exactly balanced. A *magnetic field*, $\vec{B}$, famously has the property that $\vec{\nabla} \cdot \vec{B} = 0$. Fields of this type all called *circumferential* or *solenoidal*.

Example 1.5

Consider the vector field: $\vec{A} = h(x,y)\,\hat{x} + y^2\,\hat{y}$. What are the conditions placed upon h so that $\vec{\nabla} \cdot \vec{A} = 0$?

We use Eq. (1.45) for del and then find the inner product of this with $\vec{A}$ to get

$$\vec{\nabla} \cdot \vec{A} = \frac{\partial h(x,y)}{\partial x} + \frac{\partial}{\partial y} y^2 = \frac{\partial h(x,y)}{\partial x} + 2y.$$

Setting this equal to zero gives
$$\frac{\partial h(x,y)}{\partial x} = -2y \ .$$
We see that the above is true when $h(x,y) = -2xy$.

PROBLEMS

1.16 Compute the divergence of the following vector fields:

a) $\vec{A} = \sqrt{x^2 + y^3}\ \hat{x}$
b) $\vec{B} = A_o y e^{-kx^2}\ \hat{x} + \sin(kz)\ \hat{z}$, where A and k are constants.
c) $\vec{B} = xy^{1/3}\ \hat{x} + 5z^3\ \hat{y} + 4x(y+1)\ \hat{z}$
d) $\vec{E} = 5y^2 \sinh(kx^2)\ \hat{x} + 6\cosh(x+y)\ \hat{y} + 5\cosh(2z^2)\ \hat{z}$

1.17 Let $\vec{r} = x\ \hat{x} + y\ \hat{y} + z\ \hat{z}$. Show that $\vec{\nabla} \cdot \vec{r} = 3$.

1.18 Using $\vec{r}$ as defined in the previous problem, show that $\vec{\nabla} \cdot (\vec{r}f(r)) = 3f(r) + r\frac{df}{dr}$, where, $r = |\vec{r}|$.

1.19 Show that $\vec{\nabla} \cdot f\vec{A} = (\vec{\nabla}f) \cdot \vec{A} + f\vec{\nabla} \cdot \vec{A}$. Assume that $f = f(x,y,z)$.

1.20 Let $\vec{A} = 5x^2\ \hat{x} + 6y^3\ \hat{y} - (x+z)\ \hat{z}$ and $\vec{B} = (2+x)\ \hat{x} - 2y^2\ \hat{y} + z^2\ \hat{z}$. Compute
$$\vec{\nabla} \cdot (\vec{A} \times \vec{B})\ .$$

1.21 Consider the three dimensional path differential vector $\vec{dl} = dx\ \hat{x} + dy\ \hat{y} + dz\ \hat{z}$, and some scalar function $g = g(x,y,z)$. Suppose g is *analytic* then,
$$dg = \frac{\partial g}{\partial x}dx + \frac{\partial g}{\partial y}dy + \frac{\partial g}{\partial z}dz\ .$$
Show that $\vec{\nabla}g \cdot \vec{dl} = dg$.

The Curl

It seems only natural now to consider the cross product of del with a vector field. This operation is referred to as finding the *Curl*. As with the divergence, it has an interesting and useful physical description. From its name you may have guessed that is gives a measure of the rotational or swirling nature of the vector field. In fact if $\vec{\nabla} \times \vec{A} = 0$ then we say that the vector field $\vec{A}$ is *irrotational*.

A vector field describing the motion of air molecules in a tornado would certainly have a non-zero curl while a vector field like that of Figure 1.5 would be zero. Recalling our formula for the cross product given in Eq. (1.28) we have that the curl of the vector field $\vec{A}$ is

$$\vec{\nabla} \times \vec{A} = \begin{vmatrix} \hat{x} & \hat{y} & \hat{z} \\ \frac{\partial}{\partial x} & \frac{\partial}{\partial y} & \frac{\partial}{\partial z} \\ A_x & A_y & A_z \end{vmatrix} . \tag{1.49}$$

It should be noted that the order of operation in the application of Eq. (1.49) is always such that an operator is placed to the left of the components of $\vec{A}$.

If $f = f(x, y, z)$ then by using the chain rule for derivatives, one can show that

$$\vec{\nabla} \times (f\vec{A}) = f\vec{\nabla} \times \vec{A} + (\vec{\nabla} f) \times \vec{A} . \tag{1.50}$$

Example 1.6

Suppose a velocity vector field is given by $\vec{v} = \beta(y\,\hat{x} - x\,\hat{y})$ m/s. Here a velocity is associated with every position (x, y) in the two dimensional plane. It must be that β is a constant with units of inverse seconds. Find the curl of this vector field.

We use formula (1.49):

$$\vec{\nabla} \times \vec{v} = \beta \begin{vmatrix} \hat{x} & \hat{y} & \hat{z} \\ \frac{\partial}{\partial x} & \frac{\partial}{\partial y} & \frac{\partial}{\partial z} \\ y & -x & 0 \end{vmatrix}.$$

expanding down the third column leads to

$\hat{z}\left(\frac{\partial}{\partial x}(-x) - \frac{\partial}{\partial y}y\right)\beta - \frac{\partial}{\partial z}(-x\,\hat{x} - y\,\hat{y})\beta = -2\beta\,\hat{z}.$

Since the curl of the field is non-zero it has a rotational nature to it. This can be seen by plotting some if of the vectors in this field for the case where $\beta = 1.0$ s^{-1}. Notice that the ro-

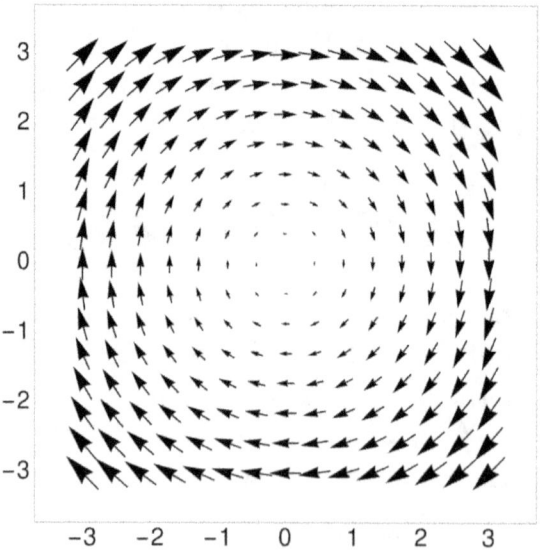

Figure 1.6: Some of the velocity vectors in the field $\vec{v} = \beta(y\,\hat{x} - x\,\hat{y})$ m/s for $\beta = 1.0$ s^{-1}.

tation in the field is clockwise and that $\vec{\nabla} \times \vec{v}$ is perpendicular to the xy plane directed into the page.

1.5. VECTOR FIELDS

PROBLEMS

1.22 Compute the curl of the following vector fields:

a) $\vec{A} = \sqrt{x^2 + y^3}\ \hat{x}$
b) $\vec{B} = A_o y e^{-kx^2}\ \hat{x} + \sin(kz)\ \hat{z}$, where A and k are constants.
c) $\vec{B} = xy^{1/3}\ \hat{x} + 5z^3\ \hat{y} + 4x(y+1)\ \hat{z}$
d) $\vec{E} = 5y^2 \sinh(kx^2)\ \hat{x} + 6\cosh(x+y)\ \hat{y} + 5\cosh(2z^2)\ \hat{z}$

1.23 Verify that

$$\vec{\nabla}(\vec{A}\cdot\vec{B}) = (\vec{B}\cdot\vec{\nabla})\vec{A} + (\vec{A}\cdot\vec{\nabla})\vec{B} + \vec{B}\times(\vec{\nabla}\times\vec{A}) + \vec{A}\times(\vec{\nabla}\times\vec{B})\ .$$

Hint: Use the BAC-CAB rule.

1.24 An *incompressible* fluid, that is, a fluid where the mass density remains constant throughout, can be described by a velocity vector field that is irrotational and is given by

$$\vec{v} = v_x(x,y)\ \hat{x} - v_y(x,y)\ \hat{y}\ .$$

Show that this fact leads to the *Cauchy-Riemann* relations:

$$\frac{\partial v_x}{\partial x} = \frac{\partial v_y}{\partial y} \quad \text{and} \quad \frac{\partial v_x}{\partial y} = -\frac{\partial v_y}{\partial x}\ .$$

1.25 Is the following vector field irrotational?

$$\vec{E} = \frac{k\ \hat{r}}{r^2},$$

Here $r = \sqrt{x^2 + y^2 + z^2}$, $\hat{r} = (x\ \hat{x} + y\ \hat{y} + z\ \hat{z})/\sqrt{x^2 + y^2 + z^2}$ and k is a constant.

Triple Products Involving $\vec{\nabla}$

Often we will have expressions where the del operator appears more than once, successively, as a factor. For example, consider

$$\vec{\nabla} \cdot \vec{\nabla} f , \qquad (1.51)$$

where f is a scalar function and $f = f(x, y, z)$. The divergence of the del operator with itself has a special name, the *Laplacian*.

$$\vec{\nabla} \cdot \vec{\nabla} = \nabla^2 = \frac{\partial^2}{\partial x^2} + \frac{\partial^2}{\partial y^2} + \frac{\partial^2}{\partial z^2} . \qquad (1.52)$$

So, Expression (1.51) can simply be written as $\nabla^2 f$. When this operation equals zero we have a *homogeneous* second order partial differential equation known as Laplace's equation:

$$\nabla^2 f = 0 . \qquad (1.53)$$

When $\nabla^2 f$ is non-zero we have a *nonhomogeneous* second order partial differential equation referred to as *Poisson's equation*:

$$\nabla^2 f = g(x, y, z) . \qquad (1.54)$$

Now, suppose we consider the curl of a gradient. That is,

$$\vec{\nabla} \times \vec{\nabla} f . \qquad (1.55)$$

It can be shown that $\vec{\nabla} \times \vec{\nabla} f = 0$. Therefore, **the curl of a gradient is exactly zero**. This then implies that the gradient of a scalar function is irrotational.

Another interesting triple product involving the del operator is the divergence of a curl:

$$\vec{\nabla} \cdot \vec{\nabla} \times \vec{A} . \qquad (1.56)$$

1.5. VECTOR FIELDS

It turns out that $\vec{\nabla} \cdot \vec{\nabla} \times \vec{A} = 0$ so that **the divergence of a curl is exactly zero.** Therefore, the curl of a vector field is circumferential.

Now, consider the curl of a curl: $\vec{\nabla} \times (\vec{\nabla} \times \vec{A})$. Using the BAC-CAB rule this can be written as

$$\vec{\nabla} \times (\vec{\nabla} \times \vec{A}) = \vec{\nabla}(\vec{\nabla} \cdot \vec{A}) - \vec{\nabla} \cdot \vec{\nabla} \vec{A} . \qquad (1.57)$$

Note, in Eq. (1.57), that one can take the gradient of a vector.

PROBLEMS

1.26 Compute the Laplacian of the following functions.

a) $f(x,y) = \sqrt{x^2 + y^3}$
b) $g(x,y,z) = A_o y e^{-kx^2} \sin(kz)$, where A and k are constants.
c) $g(x,y) = xy^{1/3} + 5z^3 + 4x(y+1)$
d) $f(x,y) = 5y^2 \sinh(kx^2) + 6\cosh(x+y)$

1.27 Verify that Expression (1.55) evaluates to zero. Assume $f = f(x,y,z)$, that the second derivatives of f exist and that the order of differentiation does not matter.

1.28 Use the BAC-CAB rule to derive Eq. (1.57).

1.29 In electrostatics, Poisson's equation relates a scalar voltage V, to a charge density ρ as

$$\nabla^2 V = -\frac{\rho}{\epsilon_o} ,$$

where ϵ_o is a constant. If $V(x,y) = 2x + 3y^2$, determine ρ.

1.30 In electrostatics, if the voltage, V, is known on the boundary of a region then the voltage within the interior is governed by Laplace's equation. That is

$$\nabla^2 V = 0 .$$

If the region is in two dimensions, what are some functions $V = V(x, y)$, which satisfy Laplace's equation?

1.6 Flux of a Vector Field

It is often useful to be able to quantify the "amount" of a vector field that is passing through a given surface. A useful definition that gives us this value is called the flux, Φ, of a vector field. Here, flux is defined slightly differently from the particle flux discussed earlier in this chapter.

Consider a vector field $\vec{B}(x, y, z)$ defined in three dimensional space. Now, suppose this vector field is passing through a surface as shown in Figure 1.7. Here, one of these field vectors at the surface is sketched. At the same point, a unit vector $\hat{n}$, normal to the surface at this point, is depicted. Both of these vectors originate from some infinitesimal area on the surface dA.

From this we define a differential area vector $d\vec{A}$. This is simply

$$d\vec{A} = \hat{n} \, dA . \tag{1.58}$$

Now, we define the differential of flux of $\vec{B}$ through the surface dA as $d\Phi = \vec{B} \cdot d\vec{A}$. Therefore, the flux through the entire surface S would be

$$\Phi = \int_S \vec{B} \cdot d\vec{A} . \tag{1.59}$$

1.6. FLUX OF A VECTOR FIELD

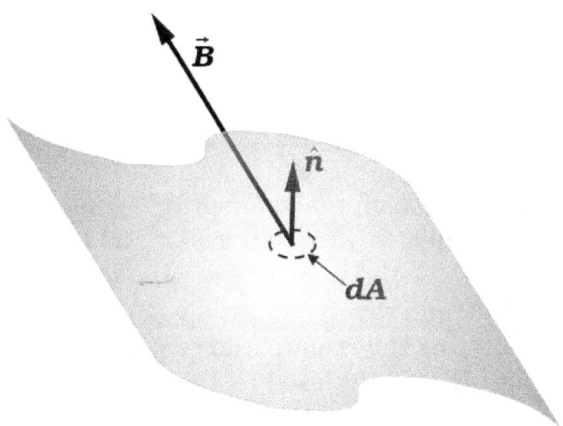

Figure 1.7: A vector from the vector field $\vec{B}$ at area infinitesimal dA on a surface S along with a unit normal vector $\hat{n}$ at the same point.

Here S denotes that the integration is over some surface S and is then actually a double integral. To apply Eq. (1.59) in the correct way, we need the area differential written for the surface in question, using the coordinates of the system involved, and accordingly $\hat{n}$ written as a unit vector normal to the surface at each dA. This may seem difficult, and for certain situations, it can be. However, often due to symmetry and the simplicity of the surface, these items can be straight forward to obtain.

Though the flux, Φ is a scalar, it can be negative or positive depending upon the orientation of the unit vector $\hat{n}$. We will use the convention that $\hat{n}$ is always directed towards the outside of a fully enclosed surface. For surfaces that do not fully enclose some volume of space, the orientation will be established beforehand.

Example 1.7

Find the flux of the vector field $\vec{C} = (3x^2 + \sin y)\,\hat{z}$ through a square on the xy plane where $0 \le x \le \pi$ and $0 \le y \le \pi$.

Here $dA = dxdy$ and $\hat{n} = \hat{z}$ so that $\vec{C} \cdot d\vec{A} = (3x^2 + \sin y)\,\hat{z} \cdot dxdy\,\hat{z}$. Performing the required inner product and inserting this result into Eq. (1.59) leads to

$$\Phi = \int_0^\pi \int_0^\pi (3x^2 + \sin y)\,dxdy\ .$$

Integrating first with respect to x leads to

$\int_0^\pi (x^3 + x\sin y)|_0^\pi\,dy = \int_0^\pi (\pi^3 + \pi \sin y)\,dy$

Now, integrate with respect to y:

$\int_0^\pi (\pi^3 + \pi \sin y)\,dy = (y\pi^3 - \pi \cos y)|_0^\pi = \pi^4 + 2\pi\ .$

An interesting physical description for the flux of a vector field comes from the velocity vector field depicted in Figure 1.5. Here, the vector $\vec{v}$ has SI units m/s. Therefore, the flux of this field through a surface must have the units of $\vec{v} \cdot d\hat{A}$ which would be (m/s) m^2 or m^3/s. That is, the flux of the velocity vector field through some surface of interest gives the volume flow rate through the surface. One can now clearly see what was meant by the phrase "the amount of a vector field that is passing through a given surface", used at the beginning of this section to describe the flux.

PROBLEMS

1.30 Find the flux of the uniform vector field $\vec{A} = 2\,\hat{x} + 3\,\hat{z}$ through a square on the xy plane of side width 2.0 in arbitrary units. Let $\hat{n} = \hat{z}$.

1.31 Find the flux of the uniform vector field $\vec{A} = 4\,\hat{y} + 3x\,\hat{z}$ through a cube of side length 3.0 in arbitrary units.

1.32 Find a unit vector normal to the plane of $y = x$. Hint: use the inner product. Use this unit vector to find the flux of $\vec{B} = 2x\,\hat{x} - 3y^2\,\hat{y}$ through a rectangle on this plane such that $0 \leq x \leq 2.0$, $0 \leq y \leq 2.0$ and $0 \leq z \leq 3.0$.

1.7 Important Integral Theorems

In this section three integral theorems that involve the del operator are discussed. These theorems, especially two of them, will be of much use to us later when we deal with Maxwell's equations. Before proceeding to our discussion of these though, it will be helpful to review the concept of a line integral.

Line Integrals

A line integral gives the distance along a line. Let an infinitesimal distance along a path be dl. Then, the distance d from a to b is simply given by

$$d = \int_a^b dl \ . \tag{1.60}$$

This works well assuming that we know how to write dl. If the path is along a Cartesian coordinate axis, say the x axis, then $dl = dx$ and then Eq. (1.60) gives us the distance along the axis from point a to b.

Often, one has a function which describes a path in two or three dimensions. In this case, we need not use Eq. (1.60) to find the distance along the designated path between two points. In fact, if $f(x)$ gives a path in the two dimensional plane, then the distance d from point a to b can be shown to be given by

$$d = \int_a^b \sqrt{1 + \left(\frac{df}{dx}\right)^2}\, dx \ . \tag{1.61}$$

A more useful line integral for the study of electromagnetic theory is the integral that gives us the total change in some function on going along a designated path from point a to b. To see how to write this line integral consider a function in three dimensional space $g = g(x, y, z)$. Then, the infinitesimal change in g over some infinitesimal path vector $\vec{dl}$ will be $\vec{\nabla} g \cdot \vec{dl}$. Therefore, the total change in g, Δg, on going from a to b is simply

$$\Delta g = \int_a^b \vec{\nabla} g \cdot \vec{dl} \ . \tag{1.62}$$

By the fundamental theorem of calculus this becomes

$$\int_a^b \vec{\nabla} g \cdot d\hat{l} = g(b) - g(a) \ . \tag{1.63}$$

One might rightfully argue that the integral in Eq. (1.63) is not terribly useful when the function g is already well known. However, it turns of that there are

1.7. IMPORTANT INTEGRAL THEOREMS

many other situations where only the gradient of a scalar function is known, that is some vector field, and an integral of the form of Eq. (1.63) needs to be computed. For example the integral

$$\int_a^b \vec{E} \cdot \vec{dl}, \qquad (1.64)$$

will need to be computed many times in this course.

So how is one to interpret the vector differential $\vec{dl}$? As one might expect, this differential represents and infinitesimal vector that point in the direction of the path. For example, suppose the path is along the x axis in a Cartesian system. Then we have that $\vec{dl} = dx\,\hat{x}$. In general for any arbitrary path in three dimensions

$$\vec{dl} = dx\,\hat{x} + dy\,\hat{y} + dz\,\hat{z}. \qquad (1.65)$$

Example 1.8

Compute $-\int_\infty^b \vec{E} \cdot \vec{dl}$, for the case where,

$$\vec{E} = \frac{K}{r^2}\,\hat{r}.$$

Here, K is a constant and r is variable position.

We have a vector field that is a function of one position coordinate so that $\vec{dl} = dr\,\hat{r}$. After computing the inner product, the integral becomes

$$-\int_\infty^b \frac{K}{r^2} dr = \frac{K}{r}\Big|_\infty^b.$$

This evaluates to

$$\frac{K}{b} - 0 = \frac{K}{b}.$$

The Fundamental Theorem for Divergences

In this section we will discuss an important integral theorem involving the divergence of a vector field. Often referred to as *Gauss's theorem*, it is perhaps easier to remember if one refers to it by the descriptive title: *The Fundamental Theorem for Divergences*. This integral theorem states that the divergence of a vector field through some volume τ is equivalent to the flux of the vector field through a surface S enclosing the volume. That is,

$$\int_\tau \vec{\nabla} \cdot \vec{A} \, d\tau = \oint_S \vec{A} \cdot d\vec{A} \, . \qquad (1.66)$$

Here, $d\tau$ is the volume differential for the coordinate system of choice and the subscript τ on the integral on the left of Eq. (1.66) implies volume integration over the volume τ so that this integration is actually a triple integral. For Cartesian coordinates, $d\tau = dxdydz$. The circle written across the integral on the right of Eq. (1.66) is to remind the reader that the integration is over a closed surface, that is, a surface that fully encloses the volume τ.

The Fundamental Theorem for Curls

Another useful integral theorem that often goes by the name of *Stoke's theorem* we will refer to as *The Fundamental Theorem for Curls*. This integral formula relates the flux of the curl of a vector field through a non-closed surface, S, to the line integral of the vector field around a closed path bounding the surface. That is

$$\int_S (\vec{\nabla} \times \vec{A}) \cdot d\vec{A} = \oint \vec{A} \cdot d\vec{l} \, . \qquad (1.67)$$

1.7. IMPORTANT INTEGRAL THEOREMS

Though Eqs. (1.66) and (1.67) may appear to be difficult to deal with, it turns out that if the vector field and the surface in question have a lot of symmetry, they can be straight forward to evaluate.

Example 1.9

Consider the vector field $\vec{A} = y\,\hat{x} + 2x\,\hat{y}$ in arbitrary units. Using $\vec{A}$ show that both sides of Eq. (1.67) yield the same result for a surface of a square of side length 1.0 lying in the xy plane where $0 \leq x \leq 1.0$ and $0 \leq y \leq 1.0$.

We work on the left side of Eq. (1.67) and begin by finding the curl of $\vec{A}$

$$\vec{\nabla} \times \vec{A} = \begin{vmatrix} \hat{x} & \hat{y} & \hat{z} \\ \frac{\partial}{\partial x} & \frac{\partial}{\partial y} & \frac{\partial}{\partial z} \\ y & 2x & 0 \end{vmatrix} = \hat{z}\left(\frac{\partial}{\partial x}2x - \frac{\partial}{\partial y}y\right) = \hat{z} \;.$$

Now, $d\vec{A} = \hat{z}\,dxdy$ so that the integral on the left of Eq. (1.67) becomes

$$\int_0^1 \int_0^1 \hat{z}\cdot\hat{z}\,dxdy = \int_0^1 \int_0^1 dxdy = 1.0 \;.$$

Now, for the integral on the right of Eq. (1.67) we have to compute the line integral around the path that bounds the square in the xy plane. We will take the counter clockwise path shown below.

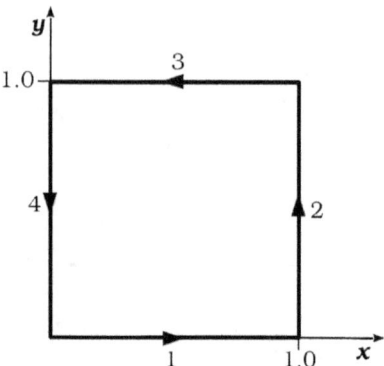

Break this into four line integrals, one for each section shown in the Figure above.

$$\overbrace{\int_0^1 (y\,\hat{x} + 2x\,\hat{y}) \cdot \hat{x}\,dx}^{1} + \overbrace{\int_0^1 (y\,\hat{x} + 2x\,\hat{y}) \cdot \hat{y}\,dy}^{2} +$$

$$\overbrace{\int_1^0 (y\,\hat{x} + 2x\,\hat{y}) \cdot (\hat{x}\,dx)}^{3} + \overbrace{\int_1^0 (y\,\hat{x} + 2x\,\hat{y}) \cdot (\hat{y}\,dy)}^{4}$$

Integrals 1 and 4 are zero since y and x are zero along these paths respectively. Computing the remaining two integrals leads to

$\int_0^1 2x\,dy + \int_1^0 y\,dx = 2 - 1 = 1$. Note how in integral 2 $x = 1$ and in integral 3, $y = 1$.

PROBLEMS

1.33 Verify Eq. (1.66) using $\vec{A} = x\,\hat{x} + y\,\hat{y}$ in arbitrary units, for the unit cube $0 \leq x \leq 1.0$, $0 \leq y \leq 1.0$ and $0 \leq z \leq 1.0$.

1.8. CYLINDRICAL COORDINATES

1.34 Often if the integral $\oint \vec{A} \cdot \vec{dl}$ is required, it is easier to compute the left side of Eq. (1.67) to get the result. Using the left side of Eq. (1.67) find $\oint \vec{A} \cdot \vec{dl}$, around the boundary of the square where $0 \leq x \leq 1.0$ and $0 \leq y \leq 1.0$, for the following vector fields:

a) $\vec{A} = 2y\,\hat{x} + z\,\hat{y}$
b) $\vec{A} = 2y\,\hat{x} + z\,\hat{y} + x^2\,\hat{z}$
c) $\vec{A} = 2y\,\hat{x} + \sin x\,\hat{y}$

1.8 Cylindrical Coordinates

Though we will rarely deal with vectors in cylindrical coordinates in this course, it will be the case that at times the need will arise to perform computations with scalar functions that are more easily described in this coordinate system. For example, functions that possess a lot of circumferential symmetry. Cylindrical coordinates are a straight forward extension of polar coordinates into a third dimension. The coordinates for such a system are depicted in Figure 1.8.

The polar system, shown on the left in Figure 1.8, gives a convenient way to denote coordinates in two dimensions for a curve that has circular symmetry. The point P has polar coordinates (r, ϕ). Using some basic trigonometry, we see that these coordinates relate to their Cartesian counterparts as

$$x = r\cos\phi, \qquad (1.68)$$

and

$$y = r\sin\phi. \qquad (1.69)$$

44   CHAPTER 1. VECTOR ANALYSIS

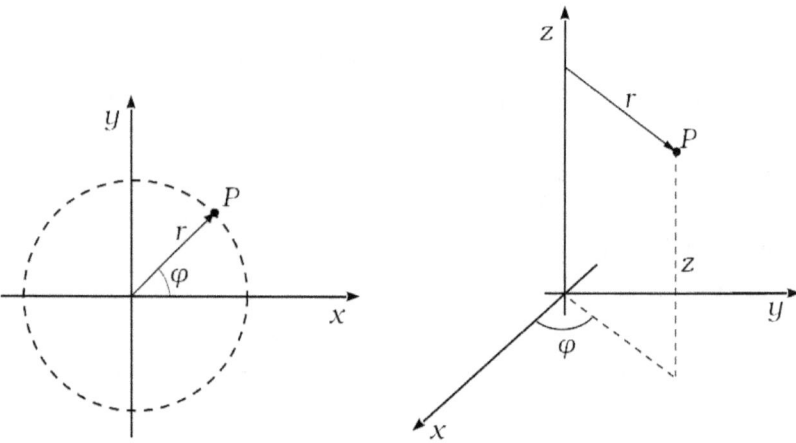

Figure 1.8:   Depiction of the two dimensional polar coordinate system on the left and the three dimensional cylindrical system on the right.

In the cylindrical system on the right of Figure 1.8 the point P is given by the cylindrical coordinates (r, ϕ, z) which relate to the Cartesian coordinates as

$$x = r \cos \phi , \tag{1.70}$$

$$y = r \sin \phi , \tag{1.71}$$

and

$$z = z . \tag{1.72}$$

It is often useful to be able to write area and volume differentials for such coordinate systems. For the polar system, using the pie shaped area shown in Figure 1.9 we see that a differential for area in the plane is simply

$$dA_{(polar)} = r dr d\phi . \tag{1.73}$$

1.8. CYLINDRICAL COORDINATES

From the volume wedge shown on the right side of Figure 1.9 it can be seen that the differential for volume in cylindrical coordinates is

$$d\tau_{(cyl.)} = r\, dr\, d\phi\, dz \ . \tag{1.74}$$

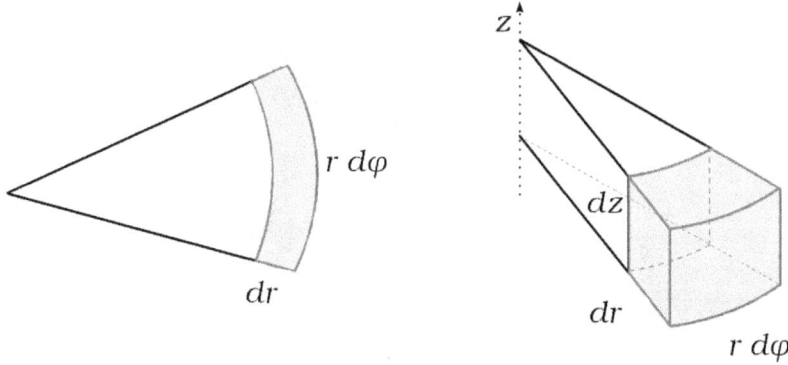

Figure 1.9: Differentials in the two dimensional polar coordinate system on the left, the three dimensional cylindrical system on the right.

For the cylindrical system one can define various area differential depending upon the surface of interest. If the surface is the outside of an upright cylinder, which is often the case, the area differential is simply

$$dA_{(cyl.)} = r\, d\phi\, dz \ . \tag{1.75}$$

One can give vectors in cylindrical coordinates in terms of basis vectors just as in the Cartesian system. for example a vector $\vec{A}$ in cylindrical coordinates would be

$$\vec{A} = A_r\, \hat{r} + A_\phi\, \hat{\phi} + A_z\, \hat{z} \ . \tag{1.76}$$

Here the components could be constants or functions of the coordinates (r, ϕ, z). To transform vector components from the Cartesian system to cylindrical one would use the transformation rules of Eqs. (170) through (172). However, more care is required when convert between basis vectors. As shown back in Figure 1.2 for spherical coordinates, basis vectors in curvilinear coordinates vary with the location of the vector. To convert these to the Cartesian system we need to use a little trigonometry to project them onto the xyz system. For example, from Figure 1.10 the cylindrical unit vector $\hat{r}$ can be written as

$$\hat{r} = \cos\phi \, \hat{x} + \sin\phi \, \hat{y} \; . \tag{1.77}$$

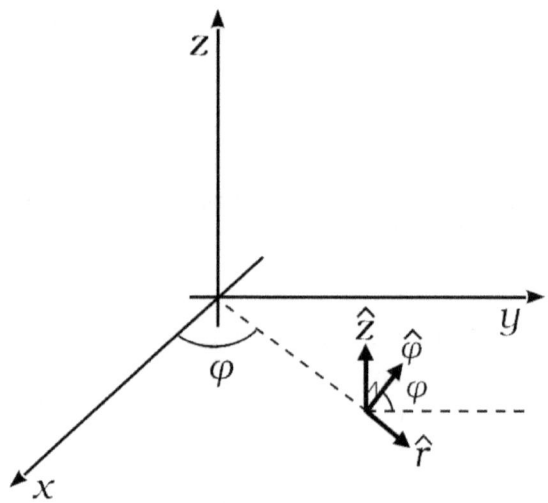

Figure 1.10: The cylindrical coordinate system with basis unit vectors denoted.

The transformation of vectors between coordinates systems is beyond the scope of this text but from the ex-

1.8. CYLINDRICAL COORDINATES

ample given above one could see how the process would unfold and the complete process for such conversions can be found in advanced texts on vector analysis.

However, we would like to cover briefly how one would arrive at the correct form for the gradient of a function in cylindrical coordinates. Recall back in Problem 1.21 that the infinitesimal path vector $\vec{dl}$ in Cartesian coordinates was given as

$$\vec{dl} = dx\,\hat{x} + dy\,\hat{y} + dz\,\hat{z}\,. \tag{1.78}$$

From Figure 1.9 we see that the analogue of this for the cylindrical system would be

$$\vec{dl} = dr\,\hat{r} + rd\phi\,\hat{\phi} + z\,\hat{z}\,. \tag{1.79}$$

Also, back in Problem 1.21 it was shown that if a function g is analytic, then its exact differential, dg could be given by $\vec{\nabla}g \cdot \vec{dl}$. Thus we would expect that for an analytic function of the cylindrical coordinates, say $f = f(r, \phi, z)$, that the following would be true

$$\vec{\nabla}f \cdot \vec{dl} = df = \frac{\partial f}{\partial r}dr + \frac{\partial f}{\partial \phi}d\phi + \frac{\partial f}{\partial z}dz\,, \tag{1.80}$$

where we have used Eq. (1.78) for $\vec{dl}$. This result gives us a route for finding the components of $\vec{\nabla}g$ in cylindrical coordinates. Using Eq. (1.79) in the left side of Eq. (1.80) leads to

$$\vec{\nabla}f \cdot \vec{dl} = (\vec{\nabla}f)_r \cdot dr\,\hat{r} + (\vec{\nabla}f)_\phi \cdot rd\phi\,\hat{\phi} + (\vec{\nabla}f)_z \cdot dz\,\hat{z}\,. \tag{1.81}$$

On comparing Eq. (1.81) with Eq. (1.80) we see that

$$(\vec{\nabla}f)_r = \tfrac{\partial f}{\partial r}\,,\quad (\vec{\nabla}f)_\phi = \tfrac{1}{r}\tfrac{\partial f}{\partial \phi}\,,\quad \text{and}\quad (\vec{\nabla}f)_z = \tfrac{\partial f}{\partial z}\,.$$

Therefore, the gradient of f in cylindrical coordinates becomes
$$\vec{\nabla} f = \frac{\partial f}{\partial r}\,\hat{r} + \frac{\partial f}{r \partial \phi}\,\hat{\phi} + \frac{\partial f}{\partial z}\,\hat{z}\ . \qquad (1.82)$$

The interesting thing about this result is that if your function f has angular symmetry, and is only a function of r, then only the first term in Eq. (1.82) is required to compute the gradient. We will take advantage of this fact several times throughout this text.

Example 1.10

Use cylindrical coordinates to derive a formula for the volume, V, of an upright cylinder of circular cross section R and length l.

We use the volume differential given in Eq. (1.74) in the appropriate triple integral:
$$V = \int_0^l \int_0^R \int_0^{2\pi} r\,d\phi\,dr\,dz\ ,$$

Integrating over ϕ, then r and finally z leads to
$$V = \int_0^l \int_0^R 2\pi r\,dr\,dz = \int_0^l 2\pi \frac{R^2}{2}\,dz = \pi R^2 l\ .$$

1.9 Spherical Coordinates

As the cylindrical system is often useful for dealing with functions of circumferential symmetry, so the spherical system is useful for dealing with situations with spherical symmetry. The spherical coordinate system has already been depicted back in Figure 1.2. In this system a scalar

1.9. SPHERICAL COORDINATES

function h can have coordinates (r, ϕ, θ). The conversion formulas from spherical to Cartesian are

$$x = r \sin\theta \cos\phi , \tag{1.83}$$

$$y = r \sin\theta \sin\phi , \tag{1.84}$$

and

$$z = r \cos\theta . \tag{1.85}$$

Forming a wedge in this system, similar to that of Figure 1.9 for the cylindrical system, we find that the differential for volume, $d\tau_{(sph.)}$, in spherical coordinates is

$$d\tau_{(sph.)} = r^2 \sin\theta dr d\theta d\phi . \tag{1.86}$$

As with the cylindrical system, we can represent a vector, $\vec{B}$, in spherical coordinates in terms of the basis vectors depicted in Figure 1.2.

$$\vec{B} = B_r \hat{r} + B_\phi \hat{\phi} + B_\theta \hat{\theta} . \tag{1.87}$$

Using the differentials from Eq. (1.86), along with these basis vectors, we can write the infinitesimal line vector in spherical coordinates as

$$\vec{dl} = dr\, \hat{r} + rd\theta\, \hat{\phi} + r\sin\theta\, \hat{\theta} . \tag{1.88}$$

As in cylindrical coordinates, a surface differential in spherical coordinates depends upon the surface in question. Frequently the surface will be of a sphere centered at the origin. In this case the area different is

$$dA_{(sph.)} = r^2 \sin\theta d\theta d\phi . \tag{1.89}$$

Following the procedure used in the previous section one can arrive at the expression for the gradient of a function $h = h(r, \phi, \theta)$,

$$\vec{\nabla} h = \frac{\partial h}{\partial r} \hat{r} + \frac{1}{r \sin\theta} \frac{\partial h}{\partial \phi} \hat{\phi} + \frac{\partial h}{r \partial \theta} \hat{\theta} . \qquad (1.90)$$

Eq. (1.90) has the same useful feature as the result for cylindrical coordinates in that if there is angular symmetry, then only the first term in the gradient need be used.

Full expressions for the gradient, divergence, curl and Laplacian in cylindrical and spherical coordinates are given in an Appendix.

PROBLEMS

1.35 Find the gradient of the following scalar functions in cylindrical coordinates:

a) $f = 2r \cos(3\phi) + 2z^4$
b) $f = 4e^{-2r\phi}$
c) $g = 2\cos(r^2\phi) + 2\cosh(r^2\phi)$
d) $g = 5r^2\phi^2 + z^2\phi^2$

1.36 Find the gradient of the following scalar functions in spherical coordinates:

a) $f = k/r^2$ for k a constant
b) $f = 4e^{-2r\phi}$
c) $g = 2\cos(r^2\phi) + 2\cosh(r^2\theta)$
d) $g = 5r^2\phi^2 + z^2\theta^2$

1.9. SPHERICAL COORDINATES

1.37 Find a vector normal (perpendicular) to the plane described by $\vec{A} = 2\,\hat{x} + 3\,\hat{y} - 4\,\hat{z}$ and $\vec{B} = -3\,\hat{y} - 4\,\hat{z}$.

1.38 A vector is said to be *normalized* when its norm equals 1.0. Let $\vec{A} = 2a\,\hat{x} + 3b\,\hat{y} - 4b^2\,\hat{z}$. What are a pair of possible values for a and b so that the norm of $\vec{A}$, equals 1.0?

1.39 An N dimensional vector space has the orthogonal basis $\{\hat{e}_1, \hat{e}_2, \hat{e}_3, \ldots \hat{e}_N\}$. Show that any vector $\vec{A}$ in this space can be given by the linear combination

$$\vec{A} = \sum_{n=0}^{N} \left(\vec{A} \cdot \hat{e}_n \right) \hat{e}_n \ .$$

1.40 Show that the Schwarz inequality is verified for the vectors $\vec{A} = 2\,\hat{x} + 3\,\hat{y} - 4\,\hat{z}$ and $\vec{B} = -3\,\hat{y} - 4\,\hat{z}$.

1.41 Use Eq. (1.37) to find the volume of the parallelepiped described by $\vec{A} = 2\,\hat{x} + 3\,\hat{y} + 4\,\hat{z}$, $\vec{B} = -3\,\hat{y} - 4\,\hat{z}$ and $\vec{C} = 2\,\hat{x} + 4\,\hat{z}$. (Assume distances are in meters.)

1.42 The *Biot-Savart Law* gives the magnetic field vector $\vec{B}$ at some point in space a distance r from a charge q moving at velocity $\vec{v}$ as

$$\vec{B} = Kq\frac{\vec{v} \times \hat{r}}{r^2} \ ,$$

where K is a constant and $\hat{r}$ is a unit vector that points from the charge to the location of the $\vec{B}$ field vector. If at the instant in question the charge is at the origin and moving at $100.0\,\hat{x}$ m/s and the $\vec{B}$ field vector is located at $(0.2, 4.0)$ meters, what is $\vec{B}$?

1.43 Show that

$$\vec{A} \times (\vec{\nabla} \times \vec{A}) = \tfrac{1}{2}\vec{\nabla}(A^2) - (\vec{A} \cdot \vec{\nabla})\vec{A} \ .$$

Let $\vec{A}$ be a two dimensional vector field.

1.44 If $\vec{A}$ and $\vec{B}$ are uniform vector fields, show that

$$\vec{\nabla}(\vec{A} \cdot \vec{B} \times \vec{r}) = \vec{A} \times \vec{B} \ ,$$

where $\vec{r} = x\,\hat{x} + y\,\hat{y} + z\,\hat{z}$.

1.45 Which of the following vector fields are irrotational? Which are circumferential?

a) $\vec{v} = 2x\,\hat{x} + 3y\,\hat{y}$
b) $\vec{B} = x^2\,\hat{x} + y^2\,\hat{y}$
c) $\vec{C} = y\,\hat{x} - x\,\hat{y} + 2\,\hat{z}$
d) $\vec{a} = (1/\sqrt{x^2 + y^2})\,\hat{x}$

1.46 Let $r = \sqrt{x^2 + y^2}$. show that

$$\vec{\nabla} \cdot \vec{\nabla} r^m = m(m+1)r^{m-2} \ .$$

1.47 Suppose the function $V = V(x)$ obeys Laplace's equation in one dimension. That is,

$$\tfrac{\partial^2 V}{\partial x^2} = 0 \ .$$

What function V makes the above true? That is, what is the general solution for Laplace's equation in one dimension?

1.48 Using a line integral, determine the distance along the line $x = 2y^2$ from $x = 0$ to $x = 4$.

1.49 Consider the vector field $\vec{E} = 2xe^{-kx}\,\hat{x}$, where k is a constant. Compute the line integral, $\int_0^\infty \vec{E} \cdot d\vec{l}$.

1.9. SPHERICAL COORDINATES

1.50 Using the two dimensional vector field $\vec{A} = A_x\,\hat{x} + A_y\,\hat{y}$, verify the fundamental theorem for divergences, Eq. (1.66). Remember, both A_x and A_y could be functions of x and y.

1.51 Using the two dimensional vector field $\vec{A} = A_x\,\hat{x} + A_y\,\hat{y}$, verify the fundamental theorem for curls, Eq. (1.67). Remember, both A_x and A_y could be functions of x and y.

1.52 Let us define a new vector operator, $\hat{C}$, called the cel operator, where

$$\hat{C} = \tfrac{\partial}{\partial y}\,\hat{x} + \tfrac{\partial}{\partial x}\,\hat{y} + \hat{z}\,.$$

Let f be the scalar function $f = f(x, y, z)$. Show that $\hat{C}\cdot\hat{C}f = \nabla^2 f + f$, where ∇^2 is the two dimensional Laplacian.

1.53 Use the differential for volume in spherical coordinates to derive the formula for the volume V of a sphere of radius R. In doing so, you will find that the angular part of this triple integral will always yields 4π when the integrand is not a function of ϕ or θ.

Chapter 2

Electrostatics

2.1 Coulomb's Law

This course is a study of that substance in nature that has come to be known as *charge*. This character, or property possessed by certain atomic particles, manifests itself in the macroscopic world via a force. It is observed that there are two distinct types of charge which are traditionally referred to as positive (+) and negative (-). Most substances hold a roughly equal number of each charge type so that they appear to be *neutral* to the observer. When an isolated solid, liquid or gas holds an excess of one charge type over the other, we say that the substance is *charged*. Typically, one always refers to the excess charge within or on a material medium as the charge held by the material and henceforth in this text it will be assumed that a non-zero charge refers to an excess charge of one type or the other.

The SI unit for charge is the Coulomb (C). For our

purposes in this text, the fundamental source of all negative charge is the electron which holds the charge -1.6×10^{-19} C. Likewise the proton is the source of all positive charge with a charge per particle of $+1.6 \times 10^{-19}$ C. With respect to charge, most material mediums can be classified as being one of two types, *conductors* or *insulators*. Roughly speaking, charge can pass freely through a conductor whereas charge cannot flow through an insulator.

In the late 18th century the French scientist Charles Coulomb (1736-1806), using a delicate apparatus, was able to quantify the nature of the force between charged particles. These results indicated that the magnitude of the force, F, between two point charges is directly proportional to the absolute values of the product of the charges,

$$F \propto |q_1 q_2|, \qquad (2.1)$$

where q_1 is the total excess charge on point charge 1 and q_2 is the total excess charge held on point charge 2.

Additionally, Coulomb found that this force magnitude was inversely proportional to the square of the distance, r, between the particles' centers. That is,

$$F \propto \frac{1}{r^2}. \qquad (2.2)$$

Using these two relations together, along with a constant of proportionality, k, leads to *Coulomb's law*:

$$F = k \frac{|q_1 q_2|}{r^2}, \qquad (2.3)$$

Coulomb also found that the force was directed along a line between the centers of the charged particles and

2.1. COULOMB'S LAW

that like charges repel and unlike charges attract. This gives us a sense of direction so that we can write an expression for the force on one charge due to the other as a vector quantity.

$$\vec{F} = k\frac{|q_1 q_2|}{r^2}\hat{r} \ . \tag{2.4}$$

Here $\hat{r}$ is a unit vector in the required direction. Forces that are described by Eq. (2.4) are referred to as *Coulomb forces*.

Since forces come in equal and opposite pairs, we always take the perspective of finding the force on one charge due to another. As an example, consider two charges fixed in space as shown in Figure 2.1. We refer to such charges as *static* charges. The force between these two charges of opposite sign would be attractive. We can write an expression for the force on q_1 due to q_2, $\vec{F}_{12}$, by using the unit vector shown in Figure (2.1), in Eq. (2.4).

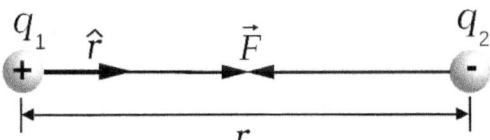

Figure 2.1: Two static charges separated by a distance r experience a Coulomb force $\vec{F}$.

Assuming the direction and notation for Cartesian coordinates we can put all of this together as

$$\vec{F}_{12} = k\frac{|q_1 q_2|}{r^2}\hat{x} \ . \tag{2.5}$$

In SI units, k takes the form:

$$k = \frac{1}{4\pi\epsilon_o}, \qquad (2.6)$$

where ϵ_o is called the *permittivity of free space*. This constant has the value $\epsilon_o = 8.85 \times 10^{-12}$ C^2/(N m^2). Its presence within our formula for Coulomb's law indicates that all of our charges reside in free space, (vacuum) or air as air very nearly has the same properties as free space when it comes to electric fields. All of our charges in this chapter can be assumed to exist in free space. We will have more to say about electric fields in material mediums in a later chapter. It should be pointed out that Eq. (2.5) is valid only when the particles are point charges or when the distance between the charged objects is much greater than their size.

It should be noted that in all cases in this text where a Coulomb force is computed, any force due to gravity is ignored. It can be shown, for any charged object we will consider, that the Coulomb force is much greater in magnitude than any gravity force involved and thus gravity forces are ignored.

Now the question arises of how to deal with the force on one fixed charge due to multiple static charges in the surrounding area? The net force acting on the charge of interest can easily be found if we assume that the *principle of superposition* holds. Here it is assumed that all of the forces between pairs of particles act independently of one another. Then, the net force, $\vec{F}$, acting on the charge of interest q_1 is given by the sum of the forces of all N charges acting acting independently upon q_1. That is,

$$\vec{F} = \vec{F}_{12} + \vec{F}_{13} + \vec{F}_{14} + \cdots + \vec{F}_{1N}. \qquad (2.7)$$

2.1. COULOMB'S LAW

We can use position vectors, and the superposition principle, to rewrite Eq. (2.5) in a more general way which gives the force on charge q_1 due to N static point charges. The required position vectors for describing the force between q_1 and q_2 are shown in Figure 2.2. Though this is a two dimensional figure the result will be valid for three dimensions as well.

Let the position vector to q_1 be $\vec{r}_1$ and the position vector to q_1 be $\vec{r}_2$. Then, a vector that points from q_2 to q_1 is $\vec{r}_1 - \vec{r}_2$. Therefore, the straight line distance between q_2 and q_1 must be

$$|\vec{r}_1 - \vec{r}_2| . \tag{2.8}$$

A position vector, $\hat{r}_{21}$, that points from q_2 to q_1 is

$$\hat{r}_{21} = \frac{\vec{r}_1 - \vec{r}_2}{|\vec{r}_1 - \vec{r}_2|} . \tag{2.9}$$

Using such vectors for all N charges in the static assembly, along with the straight line distance distance between them, Eq. (2.5) can be adapted to give the force on q_1 due to an assembly of N static charges.

$$\vec{F} = kq_1 \sum_{n=2}^{N} \frac{q_n (\vec{r}_1 - \vec{r}_n)}{|\vec{r}_1 - \vec{r}_n|^3} . \tag{2.10}$$

Example 2.1

A system of four static charges, all of identical charge q, are placed in a two dimensional system as: One at origin, one at (0,1), one at ($\sqrt{2}/2, \sqrt{2}/2$) and the final one at (1,0). Find the force on the charge at origin due to the other three. All units are SI.

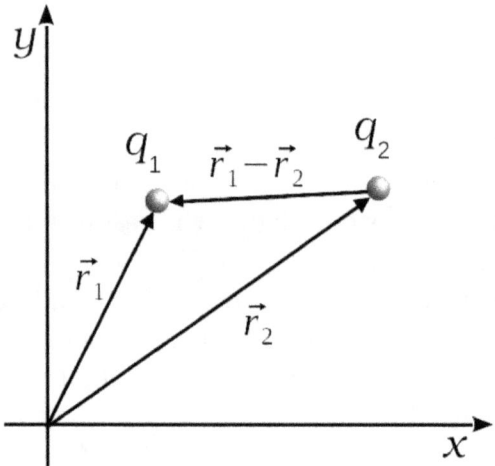

Figure 2.2: Two static charges with position vectors denoting locations.

We apply Eq. (2.10). Here $\vec{r}_1 = 0$. Assemble the other required vectors:

$\vec{r}_2 = \hat{y}$.
$\vec{r}_3 = \sqrt{2}/2 \, \hat{x} + \sqrt{2}/2 \, \hat{y}$.
$\vec{r}_4 = \hat{x}$.

Now, we use these vectors and their magnitudes in Eq. (2.10) for the case where $N = 4$:

$$\vec{F} = kq^2 \left(\frac{-\hat{y}}{1^3} - \frac{\sqrt{2}/2 \, \hat{x} + \sqrt{2}/2 \, \hat{y}}{1^3} - \frac{\hat{x}}{1^3} \right) .$$

Combining like terms leads to

$$\vec{F} = -kq^2 \left[(1 + \sqrt{2}/2) \, \hat{x} + (1 + \sqrt{2}/2) \, \hat{y} \right] .$$

2.2. THE ELECTRIC FIELD

PROBLEMS

All charges are static in the following problems.

2.1 A crude model for the helium atom has a proton of charge $+2q$ at the origin of a two dimensional system while one electron of charge $-q$ is located at $(0, d)$ and the other at $(a, 0)$. Assume all SI units. What is the force on the electron at $(0, d)$ due to the other two charged particles? What does this force become when $a \to \infty$?

2.2 Consider a point charge q placed at origin. There are an infinite number of equivalent point charges placed on the x axis to the right of this charge at positions $(n-1, 0)$ for $n = 2, 3, 4, \ldots \infty$. Show that the net force acting on the charge at the origin is

$$\vec{F} = -\frac{kq^2 \pi^2}{6} \hat{x} .$$

You will need the identity:

$$\sum_{n=2}^{\infty} \frac{1}{(n-1)^2} = \frac{\pi^2}{6} .$$

2.3 In the hydrogen atom, the distance between the proton and the electron is about 0.5 Å. Compute the Coulomb force and the gravity force for this arrangement then the ratio of these forces. Comment on the result.

2.2 The Electric Field

You may already have concluded that the procedures described in the preceding section would be tedious to apply if the number of static charges were large. This is obviously the case for most charged objects as they typically hold an enumerable number of charged atomic particles.

Much of this difficulty can be overcome by dealing with a charge density rather than a group of discrete point charges. Then, Eq. (2.10) could be replace by an integral equation.

Before we do this though it is useful to introduce the idea of an *electric field*. Often our static charge distributions will be in the form of regular arrangements that are stable. For example, a flat charged plate, a charged sphere or a straight line of charge. Since such arrangements occur so often it is useful to be able to quickly analyze how these distributions interact with other charged particles that might be in their vicinity. The electric field, which is a vector field, is a construct that enables such a procedure. The electric field is a permanent characteristic of the fixed charge distribution. With knowledge of this field, any free charge that enters into the field experiences a Coulomb force that can be easily computed with out resorting to expressions like Eq. (2.10).

To obtain an expression for the electric field due to some fixed assembly of charges, a positive test charge, $+q$, is placed in the vicinity of the distribution. The test charge will experience a Coulomb force $\vec{F}$. The electric field vector at this point, $\vec{E}$, is then defined by

$$\vec{F} = q\vec{E} . \qquad (2.11)$$

The interesting feature of Eq. (2.11) is that $\vec{E} = \vec{F}/q$ so that the electric field does not include the field due to the test charge. It is a permanent characteristic of the fixed charge distribution alone. It can be seen from Eq. (2.11) that the SI units for the electric field are N/C.

It would be a straight forward thing to modify Eq. (2.10) so that it can be used to compute the electric field

2.2. THE ELECTRIC FIELD

vector at some point, P, due to a static charge distribution. However, as mentioned earlier, it is a lot easier to deal with a density of charge rather than individual point charges. Often, the density of charge, whether on a line, surface or in a volume is known or can be reasonable estimated. In this case, the sum in Eq. (2.10) is replaced with an integral. This charge density, can be uniform or vary with position.

To illustrate the process of computing the electric field due to a static charge density, consider the electric field differential $d\vec{E}$, at some point P, due to some infinitesimal charge, dq, within some region of volume charge density ρ. This situation is depicted in Figure 2.3. Using Eq. (2.11)

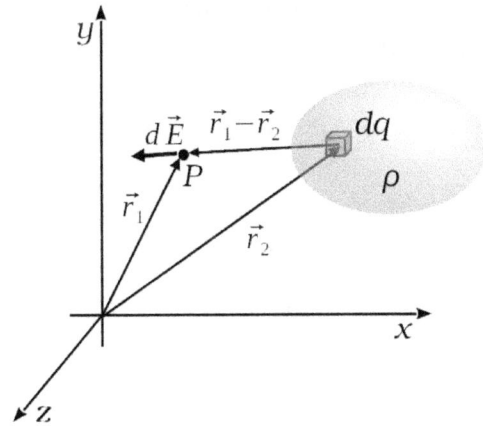

Figure 2.3: Position vectors required for finding $d\vec{E}$ at point P due to a charge distribution ρ.

in Eq. (2.10) letting $q = q_1$, $q_2 = dq$ where $dq = \rho d\tau$,

gives
$$d\vec{E} = k\frac{\rho d\tau\,(\vec{r}_1 - \vec{r}_2)}{|\vec{r}_1 - \vec{r}_2|^3}\ . \tag{2.12}$$

Integrating over the volume of the charge τ then yields the complete electric field vector at the point P.

$$\vec{E} = k\int_\tau \frac{\rho d\tau\,(\vec{r}_1 - \vec{r}_2)}{|\vec{r}_1 - \vec{r}_2|^3}\ . \tag{2.13}$$

Of course the challenging aspect of applying Eq. (2.13) is to make sure that the vectors, ρ and the differential $d\tau$, are all correct for the coordinate system involved.

Often, Eq. (2.13) can be simplified by orienting the region of charge in a favorable location in the coordinate system. When such accommodations can be made, it becomes possible to quickly write the vector $\vec{r}$, directed from dq to the point of interest, P, without resorting to finding the difference of two vectors. Letting $|\vec{r}| = r$ and $\hat{r}$ be a unit vector in the direction of $\vec{r}$ then Eq. (2.13) simplifies to

$$\vec{E} = k\int_\tau \frac{\rho d\tau}{r^2}\,\hat{r}\ . \tag{2.14}$$

To illustrate this process, let us consider a thin rod of length l which holds an excess positive charge given by a uniform linear charge density λ. Therefore, the total charge on the rod will be λl. Suppose we want to determine an expression for the electric field a distance y above the middle of this rod. This situation is depicted in Figure 2.4.

We want the differential for $\vec{E}$, $d\vec{E}$, due to the charge differential dq. This is accomplished by setting up dq, and the required vectors correctly, in Eq. (2.12). But, in

2.2. THE ELECTRIC FIELD

this situation it's easy, we need only use the differential form of Eq. (2.14) since it's straight forward to write $\hat{r}$ and r for this arrangement. Then, we simply add these differentials up across the rod to get the desired result. According to our figure, $dq = \lambda dx$. Putting it all together we get,

$$d\vec{E} = \frac{\lambda dx}{4\pi\epsilon_o(x^2+y^2)} \frac{(x\,\hat{x} + y\,\hat{y})}{\sqrt{x^2+y^2}} \,. \qquad (2.15)$$

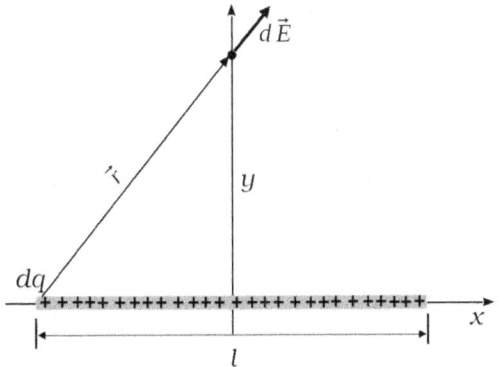

Figure 2.4: Uniform line charge of length l with uniform linear charge density λ.

Before proceeding further let's discuss the central features of Eq. (2.15). The distance between dq and q is, for any position x along the rod $\sqrt{x^2+y^2}$ so that the square of this gives (x^2+y^2). The vector on the far right side of the expression gives a unit vector along the direction of the force between dq and a test charge placed at point P, valid for any position, x, along the rod. We can now integrate both sides of Eq. (2.15) along the length of the rod and get our answer.

However, it is useful to use symmetry here to simplify the integration. Due to symmetry, all of the horizontal components of $\vec{E}$ will cancel. That is, the horizontal components on the left side will cancel those on the right. With this in mind, the $\hat{x}$ component of the unit vector in Eq. (2.15) can be ignored. We can then integrate both sides of Eq. (2.15) as

$$\vec{E} = \frac{\lambda}{4\pi\epsilon_o} \int_{-l/2}^{l/2} \frac{dx}{(x^2 + y^2)^{3/2}} \hat{y} . \qquad (2.16)$$

Notice how when $\sqrt{x^2 + y^2}$ and $(x^2 + y^2)$ are combined, we get $(x^2 + y^2)^{3/2}$. After consulting the table of integrals in an Appendix, one finds that the above leads to

$$\vec{E} = \frac{\lambda}{4\pi\epsilon_o} \frac{x}{y(x^2+y^2)^{1/2}} \bigg|_{-l/2}^{l/2} \hat{y} = \frac{\lambda l}{4\pi\epsilon_o z \left[\left(\frac{l}{y}\right)^2 + y^2\right]^{1/2}} \hat{y} .$$

It is interesting to see what happens as the distance y from the line charge gets really large. In this case, the term $(l/y)^2$ in the denominator vanishes leaving

$$\vec{E} = \frac{Q}{4\pi\epsilon_o y^2} \hat{y} , \qquad (2.17)$$

where we have let Q equal λl the total charge on the rod. This result is just what one would get a distance z from a point charge Q using Coulomb's law less the test charge. So we see that at large distances the finite line charge has an $\vec{E}$ field of a point charge Q.

2.2. THE ELECTRIC FIELD

Example 2.2

Find the electric field vector a distance z above the center of a uniformly charged disk of radius R which lies in the xy plane and is centered at the origin. The disk has area charge density σ. The situation is depicted below.

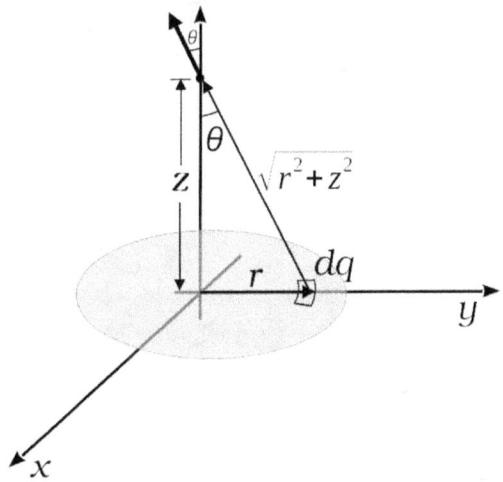

Here we use the differential for area in polar coordinates as discussed in Chapter 1. Therefore, dq will be:

$$dq = \sigma r\, dr\, d\phi\ .$$

Before going further, we note that the integral in Eq. (2.14) can be greatly simplified by taking advantage of symmetry. Notice how in the figure above as the integral sum of dq precedes around the disk the vector contributions in the plane of the disk will sum to zero. Therefore, we only require the vertical component of $d\vec{E}$ as we integrate over the disk. The vertical component is extracted from the integral when we multiply our integrand by $\cos\theta$. From the figure above we see that

$$\cos\theta = \frac{z}{\sqrt{r^2+z^2}},$$

and

$$r = \sqrt{r^2 + z^2}.$$

Putting all of this into Eq. (2.14) gives

$$\vec{E} = \frac{1}{4\pi\epsilon_0} \int_0^R \int_0^{2\pi} \frac{\sigma r dr d\phi}{(r^2+z^2)} \frac{z}{\sqrt{r^2+z^2}} \hat{z} =$$
$$\frac{1}{4\pi\epsilon_0} \int_0^R \int_0^{2\pi} \frac{\sigma z r dr d\phi}{(r^2+z^2)^{3/2}} \hat{z}.$$

The integral over ϕ yields 2π and the integral over r can be found in a table. This leads to

$$\vec{E} = \frac{z\sigma}{2\epsilon_0} \int_0^R \frac{r dr}{(r^2+z^2)^{3/2}} \hat{z} = \frac{z\sigma}{2\epsilon_0} \left[-\frac{1}{\sqrt{(r^2+z^2)}} \right] \hat{z} \bigg|_0^R =$$
$$\frac{\sigma}{2\epsilon_0} \left[1 - \frac{z}{\sqrt{(R^2+z^2)}} \right] \hat{z}.$$

PROBLEMS

All charges are static in the following problems.

2.4 A thin circular ring of radius R which holds uniform linear charge density λ lies in the xy plane centered at the origin. What is the electric field vector a distance z above the center of the loop?

2.5 For the charged disk discussed in Example 2.2, what does the final expression for $\vec{E}$ reduce to when $R \gg z$?

2.3. GAUSS'S LAW

2.6 Find the electric field vector a distance z above the center of an infinite straight line of uniform linear charge density that lies along the x axis. What does your final expression reduce to when $z \to \infty$?

2.7 A uniform surface charge density σ is on a square plate of side length l which lies in the xy plane. If the plate is centered at the origin what is the electric field vector a distance z above the center of the plate? What does your final expression reduce to when $z \to \infty$?

Even though the shapes of the charge distributions considered thus far have been favorable for the application of Eq. (2.14), one could use this formula to determined the electric field for any static charge distribution. These could be distributions with irregular shapes and cases where the charge density varies with position. However, you may have to resort to numerical techniques to evaluate the resulting integral. In the next section, we will explore a method whereby the electric field due to certain irregular static charge distributions can be easily obtained.

2.3 Gauss's Law

In this section we will consider a powerful theorem of electrostatics called *Gauss's Law*. Gauss's law relates the flux of the electric vector field through a closed surface to the static charge within the surface. More specifically, the flux of $\vec{E}$ through a closed surface is directly proportional to the total charge enclosed Q_{enc}. Mathematically speaking we have

$$\oint_S \vec{E} \cdot d\vec{A} = \frac{Q_{enc}}{\epsilon_0} . \qquad (2.18)$$

Here the constant of proportionality is the inverse of the permittivity of free space. We can learn more about Eq. (2.18) by assuming that the total charge enclosed can be given by the integral of the charge density, ρ, over the volume, τ, which the surface, S, encloses. That is,

$$Q_{enc} = \int_\tau \rho \, d\tau . \qquad (2.19)$$

Using Eq. (2.19) in Eq. (2.18) leads to

$$\oint_S \vec{E} \cdot d\vec{A} = \int_\tau \frac{\rho}{\epsilon_o} \, d\tau . \qquad (2.20)$$

Now, the fundamental theorem for divergences, from Chapter 1, can be used to convert the integral of a flux through a closed surface, on the left of Eq. (2.20), into an integral of a divergence of the electric field through the volume τ enclosed by the surface S. That is

$$\int_\tau \vec{\nabla} \cdot \vec{E} \, d\tau = \int_\tau \frac{\rho}{\epsilon_o} \, d\tau . \qquad (2.21)$$

The above equality implies that the integrands must be equivalent so that

$$\vec{\nabla} \cdot \vec{E} = \frac{\rho}{\epsilon_o} . \qquad (2.22)$$

This result is referred to as the differential form for Gauss's law while Eq. (2.18) is called the integral form.

2.3. GAUSS'S LAW

Example 2.3

An electric field in spherical coordinates is given by

$$\vec{E} = \frac{K}{r^n}\,\hat{r}\;,$$

where n is a positive integer and K is a constant with SI units appropriate for the given value of n. Use Gauss's law to find the corresponding charge density for this field.

We use Eq. (2.22) in spherical coordinates. Taking a peek at a divergence formula in an Appendix, we see that with $\vec{E}$ only a function of r, the divergence only has one term, which is

$$\vec{\nabla}\cdot\vec{E} = \frac{1}{r^2}\frac{\partial}{\partial r}r^2\left(\frac{K}{r^n}\right)\;.$$

Taking the required derivative yields

$$\vec{\nabla}\cdot\vec{E} = \frac{(2-n)K}{r^{n+1}}\;.$$

Inserting this into Eq. (2.22) and solving for ρ gives our result

$$\rho = \frac{(2-n)K\epsilon_o}{r^{n+1}}\;.$$

Note how $\rho = 0$ when $n = 2$.

PROBLEMS

2.8 An electric field in spherical coordinates is given by

$$\vec{E} = E_o e^{-\alpha r}\,\hat{r}\;,$$

where E_o and α are constants. Show that the corresponding charge density is given by

$$\rho = \epsilon_o E_o \left(\frac{2}{r} - \alpha\right) e^{-\alpha r}\;.$$

2.9 In Example 2.3 we found that $\rho = 0$ for $n = 2$. Oddly, $\vec{\nabla} \cdot \frac{K}{r^n} \hat{r} = 0$ for $n = 2$ as well. However, by the fundamental theorem for divergences it must be that

$$\int_\tau \vec{\nabla} \cdot \vec{E} \, d\tau = \oint_S \vec{E} \cdot d\vec{A} \,.$$

Show that the integral $\oint_S \frac{K}{r^2} \hat{r} \cdot d\vec{A}$ for a spherical surface of radius R centered about the origin is in fact **not** zero, but rather equals $4\pi K$. Speculate as to how $\rho = 0$ yet there is some non-zero divergence for this field. That is, what could be generating the field?

2.10 Using Coulomb's law, and the definition for the electric field, $\vec{E} = \vec{F}/q$, write the expression for the electric field due to a point charge Q placed at the origin in spherical coordinates.

In the preceding example problem, Gauss's law, in the form of Eq. (2.22), was used to find a charge density given a known electric field. Gauss's law, in the form of Eq. (2.18), is more commonly used to determine the electric field due to a known static charge distribution. The central idea in this approach is to surround some static charge distribution with a fictitious closed surface S. We call this surface a *Gaussian surface*. Since the electric vector field is involved in an integral for flux in Eq. (2.18), this method is most useful when there is a high degree of symmetry in the problem and the direction of the electric field vectors are known at the Gaussian surface. The shape of the Gaussian surface is arbitrary, it can be of any style one chooses so long as it fully encloses the charge of interest. But, as you will see, certain Gaussian surfaces are ideal for particular cases and as you gain ex-

2.3. GAUSS'S LAW

perience you will become more adept at making the best choice for a given static distribution.

To demonstrate the application of Gauss's law, we will consider a single fixed point charge q in free space. We surround the charge with a spherical shell of radius r with the point charge at its center. You may have seen this result already in practice Problem 2.10 where Eq. (2.11) and Coulomb's law were used to derive it. Everywhere on the surface of the sphere the electric field vectors will be of the same magnitude since they vary only with r. Further, since the electric field vectors are all radially directed they will all be in the same direction as $d\hat{A}$, that is, the $\hat{r}$ direction. Therefore, in this case the integrand of the integral in Eq. (2.18) becomes EdA and we get

$$\oint EdA = \frac{q}{\epsilon_o}.$$

Since E is constant over this surface it comes out of the integral and $\int dA$ simply leads to the surface area for a sphere of radius r. Eq. (2.18) now becomes

$$E(4\pi r^2) = \frac{q}{\epsilon_o}. \qquad (2.23)$$

Solving this for E then leads to the expression for the electric field vectors a distance r from a point charge q:

$$\vec{E} = \frac{q}{4\pi\epsilon_o r^2}\hat{r}. \qquad (2.24)$$

In the next example we will use Gauss's law to find a useful expression for the magnitude of the electric field between two equal and oppositely charged plates.

Example 2.4

Use Gauss's law to find an expression for the electric field strength between two parallel, infinitely large, flat, equal and opposite uniformally charged conducting plates. Let the charge on the plates be given by a constant area charge density σ.

Even though we are considering infinitely large plates, this approximation turns out to be quit reasonable for regions within the interior of the parallel plate system. Such a region is shown in the figure below.

Here, for an enclosing surface we will use what is called a *Gaussian pillbox*. Consider the left positively charged plate of a parallel plate capacitor. The plate holds area charge density σ. The pillbox is a cylinder of circular cross-section. Let each circular end of the pillbox have area A. The cylinder is then placed so that the charged plane bisects it and a selected amount of the positive charge on the plate is then enclosed within it. This setup is depicted in the figure below.

The amount of charge enclosed by the pillbox is simply σA. Some of the electric field vectors between the charge plates are shown in the figure. Notice how they are all perpendicular to the right end surface of the pillbox. This enclosing surface has three surfaces to consider. At the left end we assume that the there is no electric field outside of the parallel plate system so that the flux through this surface is zero. The second surface to consider is the side of the cylindrical pillbox. Here all of the electric field vectors are perpendicular to any unit vector that would be normal to this surface so that the inner product in Gauss's law yields zero on this entire surface. Therefore, there is no flux through this surface. Finally, the right end of the cylinder is the only surface with a non-zero flux. Here, the electric field vectors are in the same direction as a unit vector normal to this surface so that Gauss's law in this case becomes

2.3. GAUSS'S LAW

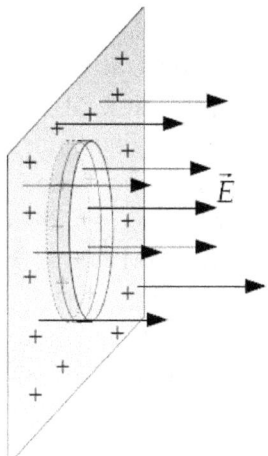

Figure 2.5: A Gaussian pillbox encloses charge on the left plate of a charged capacitor. (Right side negatively charged plate not shown.)

$$\int_S E \, dA = \frac{\sigma A}{\epsilon_o} \ .$$

E is uniform and constant so it comes outside of the integral and $\int dA$ simply yields A so that

$$EA = \frac{\sigma A}{\epsilon_o} \ .$$

Solving for E we get

$$E = \frac{\sigma}{\epsilon_o}$$

Interestingly, the electric field is constant everywhere between the plates. That is, it does not vary with position.

In concluding this section, it should be mentioned that the location of excess charge in solid materials is important for the proper application of Gauss's law. For solid conducting materials the excess charge always resides on the surface. Why is this the case? Imagine some excess positive charge placed on a solid metal sphere. Since charge is free to move around where will it end up? This then leads to another question, if the excess charge resides on the surface of a conductor, what is the electric field within its interior? On the other hand, if you encounter a situation where excess charge is uniformly distributed within a solid it must be that the object is an insulator.

Another item of note is that if an outside charge is placed within the vicinity of a conductor without making physical contact, equal and opposite charge within the conductor will flow to the surface nearest to the outside charge. We call this surface charge an *induced* charge.

PROBLEMS

2.11 A solid sphere of radius R has uniform volume charge density ρ. Use Gauss's law to find expressions for the electric field inside and outside of the sphere. Sketch a plot of E for all regions. Hint: when your Gaussian surface is within the sphere of uniform charge, it encloses some fraction of the total charge.

2.12 Use Gauss's law to find an expression for the electric field of an infinite, straight line of charge of uniform linear charge density λ. Hint: Use a Gaussian cylinder.

2.13 Use Gauss's law to find the electric field due to a single infinite flat plate of uniform charge density σ. Hint: the single

2.4. THE ELECTROSTATIC FIELD

charged plate will have non-zero field on both sides.

2.14 A solid conducting sphere of radius b has a hollow spherical cavity of radius a perfectly centered within the conducting sphere. Obviously, $a < b$. A point charge $+Q$ is fixed in place in the very center of the cavity. Use Gauss's law to find the electric field in all regions.

2.4 The Electrostatic Field

In the previous sections you were introduced to the definition of the electric field and how to arrive at mathematical expressions for this vector field for various static charge distributions. In this section we want to delve deeper into the nature of these results and in doing so we will discover certain regular characteristics for these vector fields, fields which are referred to as *electrostatic fields*.

To reveal one such property let us consider the electric field of a point charge q from the previous section.

$$\vec{E} = \frac{kq}{r^2} \hat{r}, \qquad (2.25)$$

where k is the Coulomb constant. It is of interest to consider the line integral of this $\vec{E}$ between to points r_1 and r_2 where $r_1 < r_2$. This is just

$$\int_{r_1}^{r_2} \vec{E} \cdot \vec{dl}. \qquad (2.26)$$

In the case of Eq. (2.25), $\vec{dl} = dr\hat{r}$. So that upon using

Eq. (2.25) in Eq. (2.26) we get

$$\int_{r_1}^{r_2} \frac{kq}{r^2} \hat{r} \cdot dr\hat{r} = \int_{r_1}^{r_2} \frac{kq}{r^2} dr . \qquad (2.27)$$

This evaluates to

$$\int_{r_1}^{r_2} \frac{kq}{r^2} dr = -kq \left(\frac{1}{r_2} - \frac{1}{r_1} \right) . \qquad (2.28)$$

Now, suppose rather than consider the above line integral from r_1 to r_2, we compute it around a closed circular loop of radius R where $r_1 = r_2 = R$. From eq. (2.28) we see that this evaluates to zero. That is,

$$\oint \frac{kq}{r^2} dr = 0 . \qquad (2.29)$$

From the fundamental theorem for curls this implies that

$$\int_S (\vec{\nabla} \times \vec{E}) \cdot d\vec{A} = 0 , \qquad (2.30)$$

and therefore it must be that for a fixed point charge, $\vec{\nabla} \times \vec{E} = 0$.

This is an interesting fact but it would be more useful if we could verify it for any number of fixed point charges. Recall from Eq. (2.7) that the principle of superposition holds for the Coulomb force due to multiple fixed point charges. On dividing Eq. (2.7) by a positive test charge placed at some point P near N fixed point charges we get

$$\vec{E} = \vec{E}_1 + \vec{E}_2 + \vec{E}_3 + \cdots + \vec{E}_N . \qquad (2.31)$$

Taking the curl of both sides of Eq. (2.31) gives

$$\vec{\nabla} \times \vec{E} = \vec{\nabla} \times \vec{E}_1 + \vec{\nabla} \times \vec{E}_2 + \vec{\nabla} \times \vec{E}_3 + \cdots + \vec{\nabla} \times \vec{E}_N . \qquad (2.32)$$

2.4. THE ELECTROSTATIC FIELD

Since all fields above are due to point charges, and we found that the curl of the electric field due to one point charge is zero, it must be that the right side of Eq. (2.32) equals zero. We than can state that in general, **The curl of an electrostatic field is zero.**

PROBLEMS

2.15 Which of the following fields can be electrostatic fields?

a) $\vec{E} = 2\,\hat{x} + 3y^2\,\hat{y}$ N/C
b) $\vec{E} = \sin x\,\hat{x} + 3y^2\,\hat{y}$ N/C
c) $\vec{E} = \frac{K}{r^3}\,\hat{r}$ N/C
d) $\vec{E} = 2xy^3\,\hat{x} + (y-2)x^2\,\hat{y} - 2xz^4\,\hat{z}$ N/C

2.16 Consider the following electric fields in SI units:

$$\vec{E} = \frac{K}{r^n}\,\hat{r},$$

where K is a constant and n is a positive integer. For which values of n will $\vec{E}$ be an electrostatic field?

2.17 For what values of a and b can the following be an electrostatic field?

$$\vec{E} = 2ax^2\,\hat{x} + bxy^4\,\hat{y}.$$

2.18. Let $\vec{E} = 3x\,\hat{x} - 4\,\hat{y}$ N/C. Compute the line integral

$$\int \vec{E} \cdot d\vec{l}$$

along the path of the horizontal line from $x=0$ to $x=4$ with $y=0$ and then along the vertical line from $y=0$ to $y=3$ while $x=4$.

2.19 A charge of +2.0 μC is fixed at the origin of a three dimensional Cartesian system. What is the net Coulomb force acting on this charge due to charges +3.0 μC fixed at $(0, 1, 2)$, -2.0 μC fixed at $(1, 3, -2)$ and +4.0 μC fixed at $(1, 1, 1)$.

2.20. Use Gauss's law to find the electric field inside and outside of a hollow sphere of radius R which carries a uniform surface charge density σ. Sketch a plot of the magnitude of the electric field as a function of the radial distance from the center of the sphere.

2.21. Use Gauss's law to find the electric field inside and outside of a solid sphere of radius R which carries a uniform volume charge density $\rho = \rho_o(1 - e^{-\alpha r})$, where ρ_o and α are constants. Sketch a plot of the magnitude of the electric field as a function of the radial distance from the center of the sphere.

2.22. A solid sphere of radius R has a charge density $\rho = kr$ where k is a constant and r the radial coordinate. Compute the total charge in the sphere.

2.23 For the sphere in the previous problem, use Gauss's law to find the electric field inside and outside the sphere.

2.24 Three infinite planes with surface charge density σ are configured parallel to the xz plane in a three dimensional system. One is at $y = 0$, another at $y = a$ and the third at $y = 2a$. Use Gauss's law, and the principle of superposition, to find the electric field in the regions $y < 0$, $0 < y < a$, $a < y < 2a$ and $y > 2a$.

2.4. THE ELECTROSTATIC FIELD

2.25 Find the electric field at a distance w from the origin along the axis of a uniformly charged thin hollow cylinder of radius R and length L. Let the total charge on the cylinder be Q.

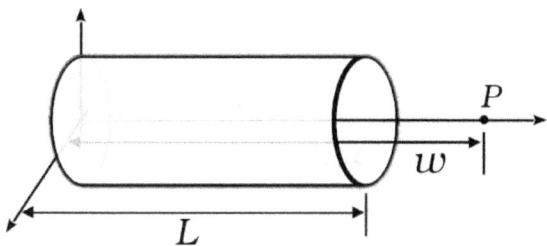

2.26 A *coaxial* cable, of arbitrary length, has a solid cylindrical conductor that holds excess charge $+Q$ and has radius a. This is surrounded by a larger thin conducting cylinder of radius b which acts as a shield for the inner conductor and holds excess charge $-Q$. Use Gauss's law to find expressions for the electric field in the regions $r < a$, $a \leq r \leq b$ and $r > b$.

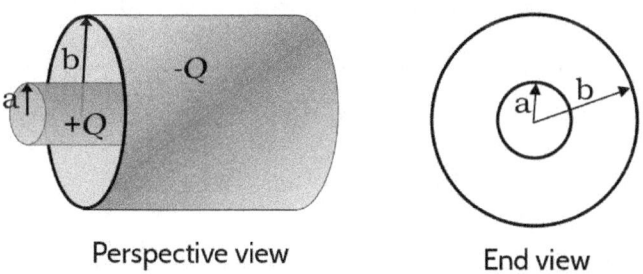

Perspective view   End view

2.27 Is $\vec{E} = 2x^2\,\hat{x} + (y+2)^3\,\hat{y}$ N/C an electrostatic field? If so, find the corresponding charge density ρ.

2.28 A thin semicircular wire holds uniform linear charge density λ. This section of wire is shown in a coordinate system below. What is the electric field at the origin?

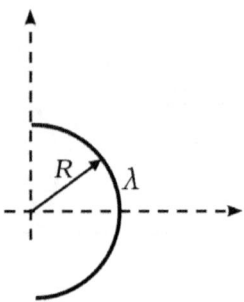

2.29 An electrostatic field is $\vec{E} = 3\,\hat{x} + 2y\,\hat{y}$. Compute the line integral

$$\oint \vec{E} \cdot d\vec{l}$$

around the closed path of the horizontal line from $x = 0$ to $x = 4$ while $y = 0$ and then along the vertical line from $y = 0$ to $y = 3$ while $x = 4$. Then along the horizontal line from $x = 4$ to $x = 0$ while $y = 3$ then back to the starting point along the vertical line $y = 3$ to $y = 0$ for $x = 0$.

2.30. A solid sphere of radius R has a charge density $\rho = k(R - r)$ where k is a constant and r the radial coordinate. Compute the total charge in the sphere. Use Gauss's law to find the electric field inside and outside the sphere.

Chapter 3

Voltage and Capacitance

3.1 Potential Energy of Static Charges

In the preceding chapter we studied a variety of static charge distributions. Since there is a force between charges, it must have required work to assemble such distributions. Since energy is conserved, assuming no energy was lost elsewhere, this work must then be stored in the static assembly of charges. One could say that the energy is stored in the electric field. However, as we will see in this section, one can compute the total stored energy by dealing only with the fixed point charges and their locations.

Recall that the curl of an electrostatic field is zero. Using Eq. (2.11) this can be written in terms of the coulomb force $\vec{F}$ as $\vec{\nabla} \times \frac{\vec{F}}{q} = 0$ which implies that

$$\vec{\nabla} \times \vec{F} = 0 \ . \qquad (3.1)$$

So the condition that the curl of the vector field $\vec{E}$ be zero implies that the curl if the conjugate force also be zero. One might recall from mechanics that a *conservative* force can be set equal to negative the gradient of a scalar potential function, U. That is

$$\vec{F} = -\vec{\nabla} U . \tag{3.2}$$

Inserting this into Eq. (3.1) we get

$$-\vec{\nabla} \times \vec{\nabla} U = 0 , \tag{3.3}$$

which we know must be true as the curl of a gradient is always zero. So $\vec{\nabla} \times \vec{F} = 0$ and $\vec{F} = -\vec{\nabla} U$ are both true for a conservative field. Now, also we see that an electrostatic field is a conservative field. There are two other mathematical relations that define conservative fields which will be discussed in a problem at the end of this chapter giving us no fewer than five ways to define a conservative vector field.

Eq. (3.2) can be used to find the potential energy between two fixed point charges. This potential energy will be equal to the work required to setup the arrangement which in turn also gives us the stored energy in the electric field. For now we will be content to have this potential in terms of the charges involved. This potential will be written in terms of the electric field later in this chapter.

Let r be the radial distance from some fixed point charge Q. Now, by Eq. (3.2), if we place a test charge, q, a distance r from Q, the force on the test charge, $\vec{F}$, must be related to the scalar potential, U, as

$$\vec{F} = -\frac{dU}{dr} \hat{r} . \tag{3.4}$$

Using Coulomb's law for F in Eq. (3.4) leads to

$$\frac{|Qq|}{4\pi\epsilon_o r^2} \hat{r} = -\frac{dU}{dr} \hat{r} . \tag{3.5}$$

Unlike with gravity, here we can have a repulsive or attractive force. For a repulsive force $dU/dr < 0$ for an attractive force

3.1. POTENTIAL ENERGY OF STATIC CHARGES

$dU/dr > 0$. These facts determine the direction, or sign, of the force. However, we want a formula for U that allows us to deal with both cases so we let the force be positive and remove the minus sign on the right of Eq. (3.5) and then straighten out things at the end.

Separating and integrating both sides from some arbitrary position r to infinity, we get

$$\int_{U(r)}^{0} dU = \frac{|Qq|}{4\pi\epsilon_o} \int_{r}^{\infty} \frac{dr}{r^2} . \qquad (3.6)$$

Integrating both sides and solving for U yields

$$U = \frac{|Qq|}{4\pi\epsilon_o r} . \qquad (3.7)$$

As was the case for gravitational potential energy, an attractive force should lead to a negative potential; that is, the particle is trapped in a well and requires kinetic energy to get out. Here, this occurs when we have two particles of opposite charge type. We should have a positive potential for a repulsive force when we have like charges. Both these conditions are meet when we write the formula for the electric potential energy of two point charges as

$$U = \frac{Qq}{4\pi\epsilon_o r} . \qquad (3.8)$$

All potential energy systems have a zero point. For Eq. (3.8), the natural zero point will be when $r \to \infty$.

A plot of Eq. (3.8) for both the repulsive and attractive situations is shown in Figure 3.1. For like charges, there is a repulsive force, $U > 0$ and the free charge q would sense a potential barrier as it approaches the fixed charge Q. For unlike charges, there is an attractive force, $U < 0$ and the free charge falls into a potential well as it moves towards Q.

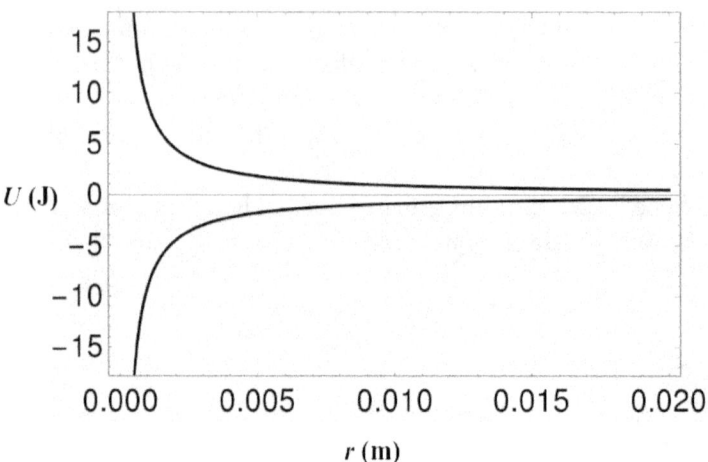

Figure 3.1: Electrical potential energy vs. distance between two point charges each of charge magnitude 1.0 μC. The upper curve is for like charges and thus a repulsive force. The lower curve is for opposite charges and an attractive force.

Eq. (3.8) can now be generalized so that it can be used to compute the electric potential energy of any number of fixed point charges. As with the Coulomb force, when multiple charges are involved it is convenient to have position vectors that give the location of each. Using a difference of position vectors to get the distance r in Eq. (3.8), we have that for N point charges

$$U = \left(\frac{1}{2}\right) \frac{1}{4\pi\epsilon_o} \sum_{i=1}^{N} \sum_{j \neq i}^{N} \frac{q_i q_j}{|\vec{r}_i - \vec{r}_j|} \ . \qquad (3.9)$$

The factor of one half was inserted since, in the double sum, the same pair of point charges will be counted twice. Eq. (3.9) may seem foreboding but as you will see in

3.1. POTENTIAL ENERGY OF STATIC CHARGES

situations where there are only a few point charges, say three, it is easy enough to apply.

Example 3.1

Consider the three point charges shown below fixed within the two dimensional plane. Compute the total electric potential energy of the configuration.

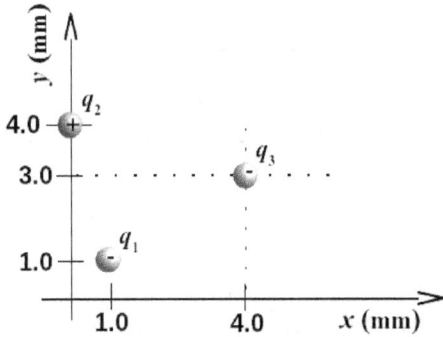

Figure 3.2: Three fixed point charges. $q_1 = $ -10.0 μC, $q_2 = $ +5.0 μC and $q_3 = $ -10.0 μC.

We have three point charges so we will have three terms to compute in Eq. (3.9). First, get the distance between the pairs.

The distance between q_1 and q_2 is

$$r_{12} = \sqrt{1^2 + (4-1)^2} = \sqrt{10} \text{ mm} .$$

Likewise for the other two pairs,

$$r_{13} = \sqrt{(3-1)^2 + (4-1)^2} = \sqrt{13} \text{ mm},$$
$$r_{23} = \sqrt{(4)^2 + (4-3)^2} = \sqrt{17} \text{ mm}.$$

We'll not use the factor of 1/2 in Eq. (13.27) since we are careful to count the pair interactions only once. Putting it all together in SI units leads to

$$U = (9.0 \times 10^9) \left[\frac{(-10 \times 10^{-6})(5.0 \times 10^{-6})}{\sqrt{10} \times 10^{-3}} + \frac{(-10.0 \times 10^{-6})(-10.0 \times 10^{-6})}{\sqrt{13} \times 10^{-3}} + \frac{(5.0 \times 10^{-6})(-10.0 \times 10^{-6})}{\sqrt{13} \times 10^{-3}} \right].$$

Using rules for exponents this can be simplified to

$$U = \left(\frac{(-10)(5.0)}{\sqrt{10}} + \frac{(-10.0)(-10.0)}{\sqrt{13}} + \frac{(5.0)(-10.0)}{\sqrt{13}} \right) \simeq -2.07 \text{ J}.$$

Example 3.2

Recall from mechanics that the work, W, by a force $\vec{F}$ in moving an object from point a to b can be given by $\int_a^b \vec{F} \cdot \vec{dl}$. Show that the work done by the Coulomb force on moving a point charge Q initially a distance r_a from point charge q to a new position a distance r_b from point charge q is equivalent to $-\Delta U$ as given by Eq. (3.8).

Here we move along the radial line so that $\vec{dl} = dr\hat{r}$. Inserting this and Coulomb's law into our integral for work gives

$$W = \int_{r_a}^{r_b} \frac{kQq}{r^2} \hat{r} \cdot dr \, \hat{r} = kQq \int_{r_a}^{r_b} \frac{1}{r^2} dr.$$

Carrying out the integration we get

$$W = -kQq \left(\frac{1}{r_b} - \frac{1}{r_a} \right).$$

Now, using Eq. (3.8) we find

$$-\Delta U = -kQq \left(\frac{1}{r_b} - \frac{1}{r_a} \right).$$

PROBLEMS

3.1 Consider the three fixed charges shown in the figure below.

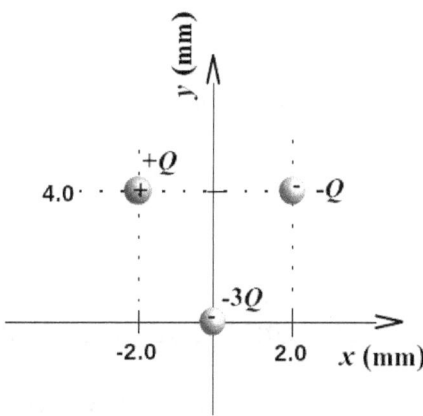

Find the total potential energy of this distribution.

3.2. The Lennard-Jones potential. A potential energy formula commonly used to describe the interaction between certain atoms and molecules is the *Lennard-Jones potential*:

$$U(r) = 4\epsilon \left[\left(\frac{\sigma}{r}\right)^{12} - \left(\frac{\sigma}{r}\right)^{6} \right].$$

where ϵ and σ are constants. The potential acts through one radial coordinate r just as our Coulomb force potential energy studied in this chapter, that is, it is a one dimensional potential. Such potentials are commonly referred to as *central potentials*. Determine the conservative force associated with this potential.

CHAPTER 3. VOLTAGE AND CAPACITANCE

3.3 The Yukawa potential. A potential energy formula commonly used to describe the interaction between nuclear particles is the *Yukawa potential*:

$$U(r) = -\frac{A}{r}e^{-rB} .$$

Here A and B are constants. As with the Coulomb force potential, it is a central potential and a function of one coordinate r. Determine the conservative force associated with this potential.

3.2 Electric Potential Difference

An electrostatic vector field, also a conservative field, can be written as the negative of the gradient of a scalar potential function. In this case, we call the scalar potential function the *electric potential difference*, or *voltage*, V, thus

$$\vec{E} = -\vec{\nabla}V . \tag{3.10}$$

The units for V are the Joule per Coulomb (J/C). This quantity is commonly referred to as voltage and we define the Joule per Coulomb to be the *volt* (V). As with all scalar potential functions, which are conjugate to some conservative vector field, they are often easier to deal with than the original vector fields when performing certain calculations.

If the electric field is known for a certain static charge distribution then the voltage of the distribution can be computed as well. To arrive at a useful formula for this

3.2. ELECTRIC POTENTIAL DIFFERENCE

consider taking the line integral of both sides of Eq. (3.10).

$$\int_a^b \vec{E} \cdot \vec{dl} = - \int_a^b \vec{\nabla} V \cdot \vec{dl} . \qquad (3.11)$$

Now, recall from Chapter 1 that $\vec{\nabla} V \cdot \vec{dl} = dV$ so that the above becomes

$$\int_a^b \vec{E} \cdot \vec{dl} = - \int_a^b dV . \qquad (3.12)$$

The integral on the right evaluates to $V_b - V_a$. As with potential energies, the voltage will always be a difference between one location and another. As you will see, there is usually a convenient zero point. For static point charge distributions, this will be at an infinite distance from the distribution. Either way, it is common practice just to write this difference as $V = V_b - V_a$ so that

$$V = - \int_a^b \vec{E} \cdot \vec{dl} . \qquad (3.13)$$

Since in the previous chapter we spent a lot of time learning how to find electric fields for various static distribution, we can now use Eq. (3.13) to find the voltage in all regions around the fixed charges. Knowledge of the voltage turns out to be very useful as it gives one a direct way to determine the work required on moving a free charge from one location to another. It is also a quantity that is relatively easy to measure in the laboratory whereas electric field magnitudes and potential energies are not.

To illustrate the use of Eq. (3.13), in finding the voltage of a distribution, let us consider the charged thin spherical shell of radius R holding uniform surface charge density σ from Problem 2.20. We seek the voltage at the center of the sphere relative to infinity. We require the electric field in all regions. From Problem 2.20 we found that $\vec{E} = 0$ for $r < R$ and $\vec{E} = (kQ/r^2)\,\hat{r}$ for $r \geq R$. Here $Q = 4\pi R^2 \sigma$. We then set up Eq. (3.13) in two regions one for $\infty > r \geq R$ and another for $0 \leq r < R$. due to symmetry we can move along the radial path and thus $\vec{dl} = dr\,\hat{r}$.

$$V = -\int_R^0 0\,dr - \int_\infty^R \frac{kQ}{r^2}\,dr\ . \qquad (3.14)$$

performing the required integration leads to

$$V = \frac{kQ}{R} = \frac{Q}{4\pi\epsilon_o R}\ . \qquad (3.15)$$

Notice how even though the electric field is zero at the center of the sphere, the voltage is not.

We see how an electric field obtained via Gauss's law can be used to compute a voltage, but does the voltage have a more direct connection to Gauss's law? The answer is yes and to see this let us consider the differential form for Gauss's law mentioned in the previous chapter. That is, $\vec{\nabla}\cdot\vec{E} = \rho/\epsilon_o$. Using Eq. (3.10) in the left side of this equation leads to

$$\vec{\nabla}\cdot -\vec{\nabla} V = -\nabla^2 V\ . \qquad (3.16)$$

Re-inserting this into Gauss's law gives Poisson's equation for the voltage

$$\nabla^2 V = \frac{\rho}{\epsilon_o}\ . \qquad (3.17)$$

3.2. ELECTRIC POTENTIAL DIFFERENCE

In a region where there is no charge, this reduces to Laplace's equation.

$$\nabla^2 V = 0 . \qquad (3.18)$$

Both Eqs. (3.17) and (3.18) are examples of partial differential equations. To find the complete solution *boundary conditions* are required. That is, if the voltage is known on the boundaries of a charge free region then Laplace's equation will yield the voltage at all points within the region.

Example 3.3

Use Laplace's equation to find an expression for V between two infinity large parallel conducting plates each with uniform surface charge density of equal and opposite total charge.

Such a system is depicted in the figure below.

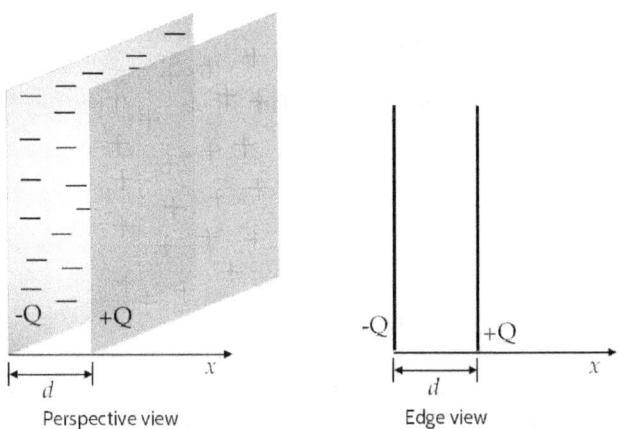

Perspective view   Edge view

Here, only a selected area of the plates are shown. However, the approximation of an infinite plate system is good so long as our interest is restricted to the interior of the region between the plates.

Since the plates are infinite this becomes just a one dimensional problem, that is $V = V(x)$. In this case Eq. (3.18) becomes an ordinary differential equation

$$\frac{d^2V}{dx^2} = 0 \ .$$

A function that obeys the above second order differential equation is

$$V = Ax + B \ ,$$

where A and B are constants. This is the so-called general solution. We need the *particular solution*. That is, the solution for our particular problem. To arrive at the particular solution for Laplace's equation we need boundary conditions. For the parallel plate system it is custom to define

$$V(0) = 0 \quad \text{(i)},$$

at the negatively charged plate. At the positive plate we let

$$V(d) = V_o \quad \text{(ii)},$$

where d is the distance between the plates. Applying boundary condition (i) we get

$$0 = A(0) + B \quad \text{so that} \quad B = 0 \ .$$

At boundry condition (ii) we have

$$V(d) = V_o = Ad \ ,$$

so that $A = V_o/d$ and then our particular solution for the parallel plate system is $V = \frac{V_o}{d}x$. So, the voltage is linear in x starting at zero on the negative plate and increasing linearly to V_o at the positive side.

3.2. ELECTRIC POTENTIAL DIFFERENCE

Using Eq. (3.10) we find that $\vec{E} = -\frac{V_o}{d}\hat{x}$.

So we see that the electric field between the plates is a uniform field with horizontal field vectors which are directed from the positive to negative plates. The magnitude of the electric field E will then be

$$E = \frac{V_o}{d},$$

which turns out to be a very useful formula for dealing with parallel plate systems. However, we can go further here since we found in Example 2.4 that the magnitude of the field between the plates was σ/ϵ_o where σ is the area charge density so that

$$V_o = \frac{\sigma d}{\epsilon_o}.$$

Also note from $E = \frac{V_o}{d}$, that the SI units for the electric field can also be given as the volt per meter (V/m).

PROBLEMS

3.4 A proton, initially at rest, is accelerated by an electric potential difference of 1.0 V. What is the final kinetic energy of the proton? This is a special unit of energy called the electron volt (eV). Hint: The electric potential *energy*, U, is related to the voltage as $U = Vq$ where q is proton charge.

3.5 A proton is accelerated from rest by a voltage of 1.0 kV over a distance of 1.0 mm. What is the speed of the proton after this acceleration?

3.6 Consider a voltage in spherical coordinates given by

$$V(r) = \frac{ke^{-\alpha r}}{r},$$

where k and α are constants. Find the charge density ρ and the total charge Q.

3.7 A parallel plate system holds a voltage of 5.0 V. If the distance between the plates is 1.0 mm, what is the surface charge density? What is the total positive charge on a 2.0 mm^2 area of the positive plate?

Looking back at Eqs. (3.4) through (3.7) we can, just by replacing symbols, find the expression for the voltage a distance r from a point charge Q. This is simply

$$V = \frac{Q}{4\pi\epsilon_o r} . \qquad (3.19)$$

Note how the voltage becomes more positive as one moves closer to positive charge while it becomes more negative as one moves towards negative charge. When plotted in each case, one gets situations similar to those found for the electric potential energy of two point charges which was plotted in Figure 3.1.

Since the electric field, due to each charge, is assumed to obey the superposition principle, then by Eq. (3.10) so must the individual voltages. On comparing Eq. (3.19) to Eq. (3.9) we see that the total voltage at some location r from N fixed point charges is given by

$$V = \frac{1}{4\pi\epsilon_o} \sum_{i=1}^{N} \frac{q_i}{|\vec{r}_p - \vec{r}_i|} , \qquad (3.20)$$

where $\vec{r}_p$ is a position vector to the observation point P and $\vec{r}_i$ is a position vector to the ith charge q_i.

3.2. ELECTRIC POTENTIAL DIFFERENCE

The above can be easily generalized for some volume charge density ρ as

$$V = \frac{1}{4\pi\epsilon_o} \int_\tau \frac{\rho\, d\tau}{|\vec{r}_p - \vec{r}_i|}, \qquad (3.21)$$

where, as usual, the challenging part is to describe both $d\tau$ and $|\vec{r}_p - \vec{r}_i|$ in terms of the configuration and coordinate system involved. In eq. (3.21), ρ could be a constant or a function of position. The volume τ can be any so long as it includes all of ρ.

Example 3.4

Use Eq. (3.21) to find the voltage a distance z above the center of the charge circular thin disk considered in Example 2.2.

Referring back to the figure in Example 2.2 we see that the distance from an element dq on the surface of the disk to a point on the z axis is $\sqrt{r^2 + z^2}$. We use the area version of Eq. (3.21):

$$V = k \int_0^R \int_0^{2\pi} \frac{\sigma r\, dr\, d\phi}{\sqrt{r^2 + z^2}},$$

where k is the Coulomb constant. As before, the angular part yields 2π. Using, a formula from the Table of integrals in an Appendix we get

$$V = 2\pi k\sigma \sqrt{R^2 + z^2}\Big|_0^R = \frac{\sigma}{2\epsilon_o}\left(\sqrt{R^2 + z^2} - z\right).$$

We can check this to see if when used in Eq. (3.10) it yields the correct expression for $\vec{E}$.

$$\vec{E} = -\vec{\nabla}V = -\frac{d}{dz}\left[\frac{\sigma}{2\epsilon_o}\left(\sqrt{R^2+z^2}-z\right)\right]\hat{z} =$$

$$\frac{\sigma}{2\epsilon_o}\left[1-\frac{z}{\sqrt{R^2+z^2}}\right]\hat{z}\,.$$

This agrees with the result found in Example 2.2.

PROBLEMS

3.8 Find the voltage at the position (0, 5) for the fixed charge distribution given in Example 3.1.

3.9 For the fixed charge distribution given in Problem 3.1, find the voltage for any position along the positive vertical axis, that is at $(0, y)$ for $y > 0$. Then, use Eq. (3.10) to find the expression for the electric field along along the positive y axis. Hint: Let the horizontal coordinates of the two charges not at $x = 0$ be $-x$ and x so that you have $V = V(x, y)$ for finding the gradient.

3.10 Find the voltage a distance z above the center of the thin ring of charge considered in Problem 2.4. Then, check your answer with Eq. (3.10).

3.3 Capacitance

In computing the voltage for the charge spherical shell and the charged disk, in the previous section, one ends up with the voltage being directly proportional to the total charge. This is often the case and the resulting constant of proportionality is very descriptive of the physical

3.3. CAPACITANCE

character of the charge arrangement. It is the custom to define the constant as

$$V = \left(\frac{1}{C}\right) Q , \qquad (3.22)$$

where C is called the capacitance and has the SI unit the Farad (F). Previously in this chapter we found that for a charged spherical shell of radius R the voltage in the center was given as

$$V = \frac{Q}{4\pi\epsilon_o R} , \qquad (3.23)$$

where Q is the total charge on the sphere. On comparing Eq. (3.22) with Eq. (3.23) we see that the capacitance for this arrangement is

$$C = 4\pi\epsilon_o R . \qquad (3.24)$$

As can be seen from this result, capacitance does not depend upon the charge or the voltage, only upon the geometry of the charged object and the medium in which it resides, in this case, free space. This is the neat thing about capacitance; its value is a fixed property of the object, or objects, that may become charged. One can think of the capacitance as a measure of the ability of the arrangement to store electric potential energy. In fact, as you will see in the next section, one can compute the stored potential energy in a charged parallel plate system given only the capacitance and the maximum voltage.

Let us now find the capacitance for the parallel plate system studied in Example 3.3. We need an expression for maximum voltage, V, relative to the zero point of the system. This is given in $V = E/d$. Now, we insert into this the known expression for the electric field magnitude between the plates

$$V = \frac{\sigma d}{\epsilon_o} . \qquad (3.25)$$

Though we originally analyzed this system for infinitely large charged plates, our results describe plates of finite size reasonably well. We therefore consider the plates to have some total area A. So, the total charge on one plate is $Q = \sigma A$ so that Eq. (3.25) becomes

$$V = \frac{Qd}{A\epsilon_o} . \qquad (3.26)$$

On comparing this with Eq. (3.22) we find that

$$C = \frac{A\epsilon_o}{d} . \qquad (3.27)$$

Notice how, as with the charged hollow sphere, the capacitance depends upon the geometry of the charge distribution and the material medium in which is resides.

The charged parallel plate system as described here is commonly referred to as a *capacitor*.

It is useful to have a formula for the total energy stored in the capacitor in terms of variables we have been dealing with in this chapter. One might be tempted to say that the total energy stored is simply VQ where Q is the total excess charge on one of the plates. However, this would be incorrect since the voltage of the parallel plate system varied as the charge was assembled, that is $V = V(q)$, where $0 \leq q \leq Q$. Therefore, the expression required to determine the stored energy, U, would be

$$dU = V(q)dq . \qquad (3.28)$$

3.4. ENERGY AND CHARGE DENSITY

Using Eq. (3.22) in this yields

$$U = \int_0^Q \frac{q}{C} dq . \qquad (3.29)$$

Integrating leads to the energy stored in the field in terms of the capacitance and total charge:

$$U = \frac{1}{2}\left(\frac{Q^2}{C}\right) . \qquad (3.30)$$

Using the fact that $Q = CV$ this can also be written as

$$U = \frac{1}{2}CV^2 . \qquad (3.31)$$

PROBLEMS

3.11 Find the capacitance for the uniformally charged sphere considered in Problem 2.11. Hint: You will first need to use Eq. (3.13) to find the voltage at the center.

3.12 Find the capacitance per unit length for the coaxial cable considered in Problem 2.26.

3.4 Energy and Charge Density

Previously in this chapter we derived a formula, Eq. (3.9), which gives the total potential energy of a static charge configuration. This is a quantity of interest but as one might suppose this formual might be awkward to use for a system of a large number of charged particles, say on the order of 10^{23}. It would then be useful to utilize a charge density within this expression rather than a point-wise sum.

A convenient way to convert Eq. (3.9) to an expression involving integration rather than summation is to first compare it with Eq. (3.20) which gives the voltage, V, at some point due to a point charge distribution. On comparing the two equations we see that Eq. (3.20) can be rewritten to give the voltage at the point of q_i due to all of the other charges as

$$V = \frac{1}{4\pi\epsilon_o} \sum_{j=1}^{N} \frac{q_j}{|\vec{r}_i - \vec{r}_j|} \ . \tag{3.32}$$

On comparing this with Eq. (3.9) we find that

$$U = \left(\frac{1}{2}\right) \sum_{i=1}^{N} V q_i \ , \tag{3.33}$$

Using a volume current density, ρ, the sum can be replaced with an integral so that

$$U = \frac{1}{2} \int_\tau V \rho \, d\tau \ . \tag{3.34}$$

What is the volume τ? Any volume, so long as it encompasses the entire charge distribution. Analogous formulas can be written for surface and line charge densities.

Example 3.5

Find the energy stored in a uniformally charge sphere of radius R.

From Problem 3.11 you should of found that

$$V = \frac{R^2 \rho}{2\epsilon_o},$$

at the center of the sphere. However, notice that in Eq. (3.32), V involves both $\vec{r}_i$ and $\vec{r}_j$ so that the V that appears in Eq. (3.34) must be a function of position. So we need to get the voltage as a function of r, where $r < R$ for our uniformally charge sphere. The electric fields inside and outside this sphere were found in Problem 2.11. Using Eq. (3.13) we get

$$V(r) = -\frac{\tau_s \rho}{4\pi\epsilon_o R^3} \int_R^r r \, dr - \frac{\tau_s \rho}{4\pi\epsilon_o} \int_\infty^R \frac{1}{r^2} \, dr,$$

where $\tau_s = (4/3)\pi R^3$, and ρ is the uniform charge density. Integrating the above and simplifying the result leads to

$$V = \frac{\tau_s \rho k}{2R}\left(3 - \frac{r^2}{R^2}\right),$$

where we have set $k = 1/(4\pi\epsilon_o)$. Inserting this into Eq. (3.34) gives

$$U = \frac{\tau_s \rho^2 k}{4R} \int_0^R \int_0^\pi \int_0^{2\pi} \left(3 - \frac{r^2}{R^2}\right) r^2 \, dr \sin\theta \, d\theta \, d\phi.$$

As usual, the angular part of the integral yields 4π and the above becomes

$$u = \frac{\tau_s \rho^2 k \pi}{R} \left(\int_0^R 3r^2 \, dr - \int_0^R \frac{r^4}{R^2} \, dr\right) = \frac{\tau_s \rho^2 k \pi}{R} \left(\frac{4}{5} R^3\right).$$

Letting $\rho = \frac{q}{\frac{4}{3}\pi R^3}$ this simplifies to

$$U = \frac{1}{4\pi\epsilon_o} \left(\frac{3}{5} \frac{q^2}{R}\right).$$

3.5 Energy in Electric Fields

In the previous section we learned how to compute the energy stored in a static distribution using the charge density and the voltage. It is useful to be able to write an expression for this energy in terms of the electric field magnitude alone. Gauuss's law can be used to eliminate V and ρ from Eq. (3.34). That is $\epsilon_o \vec{\nabla} \cdot \vec{E} = \rho$. Using this in Eq. (3.34) gives

$$U = \frac{\epsilon_o}{2} \int_\tau (\vec{\nabla} \cdot \vec{E}) V \, d\tau . \tag{3.35}$$

From a vector identity in an Appendix we get $\vec{\nabla} \cdot (\vec{E} V) = (\vec{\nabla} \cdot \vec{E}) V + \vec{E} \cdot (\vec{\nabla} V)$. Using this identity in Eq. (3.35) leads to

$$U = \frac{\epsilon_o}{2} \int_\tau \vec{\nabla} \cdot (\vec{E} V) \, d\tau - \frac{\epsilon_o}{2} \int_\tau \vec{E} \cdot \vec{\nabla} V \, d\tau . \tag{3.36}$$

In the first integral in the above, we use the fundamental theorem for divergences to convert this to a flux of $\vec{E} V$ through some surface S that encloses the volume τ. In the second integral we observe that $\vec{\nabla} V = -\vec{E}$ and we then let $\vec{E} \cdot \vec{E} = E^2$, so that Eq. (3.36) can be rewritten as

$$U = \frac{\epsilon_o}{2} \int_S (\vec{E} V) \cdot d\vec{A} + \frac{\epsilon_o}{2} \int_\tau E^2 \, d\tau . \tag{3.37}$$

Now the question becomes, which volume τ should we use? Any volume will do so long as the surface S fully encloses it. If we select τ to be *all space* then the surface integral in Eq. (3.37) will go to zero while the volume

3.5. ENERGY IN ELECTRIC FIELDS

integral will remain finite leaving the useful result

$$U = \frac{\epsilon_o}{2} \int_{All\ space} E^2\ d\tau\ . \qquad (3.38)$$

While the above limiting behavior has been assumed to be true without a formal proof, one can reason as to why this would be the case. As the volume τ is increased the volume integral can only get larger. Therefore, it must be that the surface integral decreases with increasing τ so that U remains constant. In fact, one can show that the surface integral in Eq. (3.37) goes to zero as the radius of a spherical surface S goes to infinity. Oddly, the volume integral in Eq. (3.37) goes to infinity over a radius range that includes 0. Why? Because, the energy of a point charge is infinite! This peculiar problem is usually not a detriment to the application of Eq. (3.38), if we are careful not to include the origin. Of course this is not an issue at all if our electric field is a uniform field.

If the electric field is a uniform field then E can be factored out of Eq. (3.37) leaving

$$U = \frac{\epsilon_o}{2} \int_S (\vec{E}V) \cdot d\vec{A} + \frac{\epsilon_o}{2} E^2 \int_\tau d\tau\ . \qquad (3.39)$$

Obviously, for a uniform field, $U \to \infty$ over all space but this can be made into a finite expression by dividing both sides by the volume integral which gives the finite energy density, u_o, in a uniform electric field.

$$u_o = \frac{\frac{\epsilon_o}{2} \int_S (\vec{E}V) \cdot d\vec{A}}{\int_{All\ space} d\tau} + \frac{\epsilon_o}{2} E^2\ . \qquad (3.40)$$

The first term on the right of Eq. (3.40) goes to zero so that

$$u_o = \frac{\epsilon_o E^2}{2}. \qquad (3.41)$$

PROBLEMS

3.13 Find the energy density in the uniform electric field $\vec{E} = 3.0\,\hat{x} - 4.0\,\hat{y} + 12.0\,\hat{z}$ V/m.

3.14 An electric field is given by $\vec{E} = 2x\,\hat{x} + 3y^2\,\hat{y}$ V/m. Find the energy stored in this field within the cube given by the boundaries, $0 \le x \le 4$, $0 \le y \le 4$ and $0 \le z \le 4$.

3.15 Find the energy stored in a volume charge density $\rho = \alpha r$ within a solid sphere of radius R. Hint: Use the electric fields determined Problem 2.22 in Eq. (3.38). don't forget to integrate over all space.

3.16 Consider the integrals in Eq. (3.37). Integrate both over the sphere where $0 \le r \le R$ then let $R \to \infty$. Show that in this limit the surface integral goes to zero while the volume integral diverges. Now consider the same integration over the interval $r_a \le r \le R$ where r_a is some small positive value yet not zero. Now show that when $R \to \infty$ the volume integral remains finite while the surface integral still goes to zero.

3.17 Find the voltage a distance z above the middle of a straight line charge of infinite length with uniform linear charge density λ. Then, use Eq. (3.10) to check your answer. Hint: Here V is not zero at $r \to \infty$. Therefore you must set V to be zero at some very large distance from the line charge, r_a.

3.5. ENERGY IN ELECTRIC FIELDS

3.18 A parallel plate capacitor hold a a maximum voltage of 50.0 V between plates separated by a distance of 0.2 mm. There is a uniform charge density ρ between the plates. Use Poisson's equation to find an expression for $V(x)$ between the plates.

3.19 For an electrostatic (conservative) field it must be that $\vec{\nabla} \times \vec{E} = 0$. Show that this fact leads to:

$$\frac{\partial E_z}{\partial y} = \frac{\partial E_y}{\partial z},$$
$$\frac{\partial E_z}{\partial x} = \frac{\partial E_x}{\partial z},$$
$$\text{and} \quad \frac{\partial E_y}{\partial x} = \frac{\partial E_x}{\partial y}.$$

3.20 Consider two capacitors connected in series with a voltage source and also in parallel with the same source. For capacitors in series, charge is common. For capacitors in parallel, voltage is common. What is the ratio of the stored energies of the capacitors in the series arrangement? What is the ratio of their stored energies in the parallel arrangement?

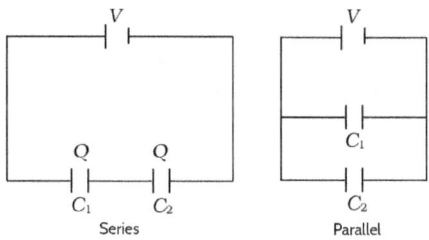

Series   Parallel

3.21 For n capacitors connected in series (or parallel), the total energy, U, stored is $U = (1/2)C_{eq}V^2$, where C_{eq} is the equivalent capacitance of the network. Use conservation of energy to show that for n capacitors connected in series,

$$C_{eq} = \frac{1}{\frac{1}{C_1} + \frac{1}{C_1} + \frac{1}{C_1} + \cdots \frac{1}{C_n}}.$$

Use conservation of energy to show that for n capacitors connected in parallel,
$$C_{eq} = C_1 + C_2 + C_3 + \cdots C_n.$$

3.22 A capacitor with air between its plates holds energy U at voltage V. If the voltage is doubled and the distance between the plates halved, what is the new energy stored?

3.23 show that for N equivalent capacitors of capacitance C in series, the equivalent capacitance is $C_{eq} = C/N$. What is C_{eq} as $N \to \infty$?

3.24 A voltage in spherical coordinates is given by
$$V(r) = \frac{Ce^{-\alpha r}}{r^2},$$
where C and α are constants. Find the corresponding $\vec{E}$ and ρ.

3.25 For a certain transmission line, the sending voltage, V, is found to obey
$$V = Ae^{-\mu x},$$
where A is an amplitude constant, μ the propagation constant and x the axial distance along the line. What is the corresponding $\vec{E}$ and ρ?

3.26 Show that the total energy stored in a uniformally charged solid hemisphere of radius R is
$$U = \frac{3}{10}\left(\frac{Q^2}{4\pi\epsilon_o R}\right),$$
where Q is the total charge in the hemisphere.

3.27 Find the total electrostatic energy stored in a uniformally charged thin spherical shell of radius R.

3.5. ENERGY IN ELECTRIC FIELDS

3.28 Find the total electrostatic energy stored in a straight line charge of length l holding uniform linear charge density λ.

3.29 A parallel plate capacitor has plate area 10.5 cm^2 and the distance between the plate is 1.25 mm. Air is between the plates. Compute the capacitance. Now, with 100.0 V applied what is the electric field between the plates, the total charge stored and the total stored energy. Then, find the force of attraction between the plates. Hint: Recall that $Work = force \times Distance$.

3.30 A flat circular ring of inner radius r_a and outer radius r_b holds a uniform surface charge density σ. If it lies in the xy plane centered at origin, what is the voltage at points along the z axis?

3.31 The Lennard-Jones (LJ) potential discussed in Problem 3.2 has a short range repulsive term which is inversely proportional to the twelfth power of the distance between the particles while the second attractive term is inversely proportional to the sixth power of the distance. The constants ϵ and σ are positive and referred to as the LJ parameters. (They are not a permittivity and a charge density). Sketch this function. This potential has a minimum at the equilibrium binding distance r_e. Find the formula for r_e.

3.32 Two additional attributes of conservative fields are that $\int_a^b \vec{E} \cdot \vec{dl}$ is independent of path and that $\oint \vec{E} \cdot \vec{dl} = 0$. Verify both of these facts for the electric field given in Problem 3.14. Let $a = 0$ and $b = 1$. First, find $\int_a^b \vec{E} \cdot \vec{dl}$ along the path $y = 0, 0 \leq x \leq 1$ then $x = 1, 0 \leq y \leq 1$. Then compute it along the straight line path from the origin to (1,1). To verify $\oint \vec{E} \cdot \vec{dl} = 0$, use any closed path.

Chapter 4

Electric Fields in Matter

4.1 Dielectrics

In the previous chapter our discussions concerning the electric field were restricted to those situations where the medium in which the field resides is free space or air. That is, we have assumed that air can be approximated as being a free space condition. In this chapter we will study what effect the presence of a material medium has upon the electric field. Since excess charge goes to the surface of a solid conductor and then by Gauss's law, the electric field in the interior is zero, we will be mainly concerned with non-conducting materials in this chapter. The motion of electrons, or *current*, in a conductor will be covered in the following chapter.

Recall that one of the important electrical properties of an insulator is that charge will not flow through it. We can think of the electrons in an insulator as all being tightly bound to the host atoms whereas in a conductor, some of the electrons

112 CHAPTER 4. ELECTRIC FIELDS IN MATTER

are free to roam. However, when an insulator is placed in an electric field, the electrons in the atoms will be pulled to one side leaving the atomic cores, that is the protons, somewhat exposed on the opposite side. This situation is depicted abstractly in Figure 4.1 where an insulating material has been placed between the charged plates of a capacitor. Cartoon drawings of some of the atoms in the insulator are shown. Notice how the electrons a pulled towards the positively charged plate.

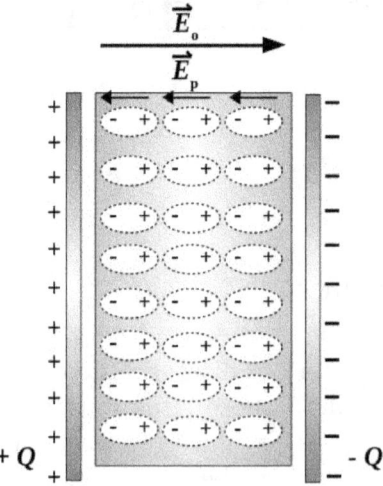

Figure 4.1: A capacitor with a dielectric medium.

When the electrons are pulled to one side in an atom by an external electric field we say the atom has become *polarized*. In Figure 4.1 the external field due to the charge plates is $\vec{E}_o$. Notice how the polarized atoms produce an electric field of their own directed opposite of the external field. We will assume the atoms produce the uniform field $\vec{E}_p$ which is

4.1. DIELECTRICS

directed opposite of $\vec{E}_o$. Then, by vector addition, the net field inside the capacitor, $\vec{E}$ is

$$\vec{E} = \vec{E}_o - \vec{E}_p \ . \tag{4.1}$$

so, the insulator acts to lower the magnitude of the electric field inside the capacitor. When an insulator is used in this capacity it is referred to as a *dielectric*. The presence of the dielectric alters the capacitance of the system and affects its ability to store energy. It is then the goal of this chapter to revisit many of the topics covered in the previous two chapters and revise our theoretical framework so as to account for the nature of electrostatic fields within dielectric materials.

The qualitative description of the behavior of a dielectric given above is informative but it is also helpful to have a quantitative description for the phenomenon of polarization. The starting point for this approach is to model the polarized atom, or molecule, as an *electric dipole*. An electric dipole is composed of two equal and oppositely charged point charges, q, separated by some distance d. It is useful to then define the dipole moment vector, $\vec{p}$, which by definition points from negative to positive charge. For example, if a negative charge, $-q$, is placed at the origin of a two dimensional system, with a positive charge $+q$ placed at position d to the right along the x axis then the dipole moment would be

$$\vec{p} = qd\ \hat{x} \ . \tag{4.2}$$

The SI units for the dipole moment will be the Coulomb-meter (Cm).

If a neutral atom is placed within a static electric field, $\vec{E}$, this field will induce a dipole moment in the atom. If the field is not too strong, it is a good first approximation to assume the the induced dipole is directly proportional to the applied field. Therefore, we can write that

$$\vec{p} = \alpha \vec{E} \ , \tag{4.3}$$

where the constant of proportionality, α, is called the *atomic polarizability*. Obviously, our dielectric material will have many of these atomic dipoles. It is then more useful to deal with the overall density of dipole moments in the material. If the number of atoms per unit volume in the dielectric is N then we define the *polarization*, $\vec{P}$, to be

$$\vec{P} = N\vec{p}. \tag{4.4}$$

Example 4.1

A parallel plate capacitor will be made having a dielectric of diamond. (A very expensive capacitor!) The voltage will be 20.0 V while the distance between the plates is 2.0 mm. Estimate the polarization for this dielectric. The atomic polarizability for the carbon atom is 1.35×10^{-20} Cm^2/V. The mass density for diamond is 3.51 g/cm^3.

Begin by using the mass density, ρ, the atomic weight for carbon, W_a, and Avogadro's number, N_A, to compute N:

$$N = \frac{N_A \rho}{W_a} = \frac{(6.02 \times 10^{23})(3.51)}{12.0} = 1.76 \times 10^{23} \text{ cm}^{-3}.$$

Now we use Eqs. (4.3) and (4.4) to get $\vec{P}$.

$$\vec{P} = (1.76 \times 10^{23})(100)^3 (1.35 \times 10^{-20}) \left(\frac{20.0}{2.0 \times 10^{-3}}\right) =$$
$$2.38 \times 10^{13} \text{ C/m}^2.$$

It should be clarified that the type polarizability being considered in this chapter is more precisely called the *electronic deformation polarizability*. There are other types of polarization that can occur which are not considered

4.2. BOUND CHARGES

in this text. If the solid in question is an ionic crystal one could have *ionic deformation polarizability* and in a molecular crystal, *orientational polarizability*. One convenient thing about the electronic deformation polarizability is that it is observed to be independent of temperature.

4.2 Bound Charges

Having defined the polarization for a dielectric we now have a convenient way to quantify the reaction of the dielectric material to an applied electric field. We would now like to study the way in which this quantity is connected with concepts introduced in the preceding chapters such as voltage, capacitance and charge density.

Fortunately, there is a way to interpret the polarization within a dielectric in terms of charge densities. Glancing at Figure 4.1 we notice how all charges all balanced by equal and opposite charges except at the left and right faces. Recall that our capacitor is assumed to be infinite so that the front, back, top and bottom edges, in a sense, do not exist. So there is a sort of bound surface charge on the outer faces of the dielectric. Let this surface charge be given by σ_b, this is referred to as a *bound surface charge density.* then it is a straight forward thing to relate it to the polarization $\vec{P}$. Notice that $\vec{P}$ has units of C/m². Then the component of $\vec{P}$ normal to the surface must give the surface charge density. Let $\hat{n}$ be a unit vector normal to the outer surface of the dielectric then we have that

$$\sigma_b = \vec{P} \cdot \hat{n} . \tag{4.5}$$

This is a great result as we know how to use a surface charge to find and electric field and a voltage.

Though often the polarization is uniform, it can vary with position within a dielectric. If so, it may be that a volume charge density, ρ_b, will develop within the dielectric. This charge density is referred to as the *bound volume charge density*. To see how this occurs consider the flux of the polarization passing through some closed surface within the dielectric. Given the units of $\vec{P}$ this flux must equal some *polarization charge* Q_p.

$$Q_p = \oint_S \vec{P} \cdot d\vec{A} . \qquad (4.6)$$

Since the dielectric was originally neutral, one must have some equal and opposite charge $-Q_p$ than remains in the region enclosed by S. Using the divergence theorem on the integral in Eq. (4.6) we then find that the charge that remains in the region τ enclosed by the surface S is

$$Q_p = -\int_\tau \vec{\nabla} \cdot \vec{P} \, d\tau . \qquad (4.7)$$

From this we conclude that

$$\rho_b = -\vec{\nabla} \cdot \vec{P} . \qquad (4.8)$$

These so-called, bound charges are sometimes referred to as *Poisson's equivalent distributions*, or *polarization charge distributions*, as they allow one to evaluate the effect of the polarization in terms of a smoothed out charge distribution. So, if the polarization is known for a given situation, one can compute the bound surface and volume charge densities.

4.2. BOUND CHARGES

It should be noted that in rare cases, certain materials can have a non-zero polarization without an external electric field being present. Such materials possess a frozen-in polarization. These materials are referred to as *electrecs*.

Example 4.2

An electrec in the shape of an upright cylinder of radius R and total length l has a frozen-in polarization $\vec{P} = kr\,\hat{z}$ in cylindrical coordinates. Here k is a constant. Determine the bound surface and volume charge.

By the term upright, we assume that the axis of the cylinder is along the z axis in our coordinate system. First we get the bound surface charge density σ_b. The unit normal vectors on the outer surface of the electrec are required. These unit vectors always point away from the polarized region. So, on the circular face of the cylinder $\hat{n} = \hat{r}$. Since $\hat{z} \cdot \hat{r} = 0$ there is no bound surface current on this surface. On the upper end surface we use Eq. (4.5) to get

$$\sigma_b = kr\hat{z} \cdot \hat{z} = kr \;,$$

and on the lower end

$$\sigma_b = kr\hat{z} \cdot -\hat{z} = -kr \;.$$

Using Eq. (4.8) we compute the bound volume charge density,

$$\rho_b = -\vec{\nabla} \cdot \vec{P} = -\tfrac{\partial}{\partial z}kr = 0 \;,$$

where we have used a formula for divergence in cylindrical coordinates from an Appendix.

Having the bound surface charge densities let us now use this result to find the electric field outside of the cylinder due to its polarization. We will find the electric field at points along the z axis for $z > l$ Letting the bottom of the cylinder be in the xy plane while the upper surface is in the plane described by $z = l$. By the principle of superposition, the net electric field along the z axis will be the sum of the electric fields due to the bound surface charge on the bottom and top faces. There are no other surface or volume charge densities to consider. Using the result from Example 2.2 we then have

$$\vec{E} = \frac{\sigma_b}{2\epsilon_o}\left[1 - \frac{z-l}{\sqrt{R^2+(z-l)^2}}\right]\hat{z} - \frac{\sigma_b}{2\epsilon_o}\left[1 - \frac{z}{\sqrt{R^2+z^2}}\right]\hat{z}\ .$$

Note that when $l \to 0$, $\vec{E} \to 0$ as the charged surfaces neutralize one another.

From the previous example, it is important to note that even though a closed Gaussian surface around the cylinder would enclose no net charge there is still an electric field in the region around the electrec. This is because Gauss's law tells use that the net flux of $\vec{E}$ through a closed surface is zero when there is no enclosed charge NOT necessarily that $\vec{E} = 0$ at the surface.

With the bound charge density known for a dielectric within a parallel plate capacitor we can give a full description to Eq. (4.1). From previous work with Gauss's law we know that the magnitude of the uniform field inside due to the plates having charge density σ will be $E_o = \sigma/\epsilon_o$. Now due to the polarization of the dielectric, the left and right faces of the dielectric act like plates with charge density σ_b so the polarization field magnitude is

4.2. BOUND CHARGES

$E_p = \sigma_b/\epsilon_o$. Using these results in Eq. (4.1) leads to

$$E = \frac{\sigma}{\epsilon_o} - \frac{\sigma_b}{\epsilon_o}. \tag{4.9}$$

PROBLEMS

4.1 A special parallel plate capacitor, as shown in Figure 4.1, has solid graphite between the plates. A carbon atom has an atomic polarizability of 1.35×10^{10} Cm^2/V. The mass density of graphite is around 2.26 g/cm^3. Assume that the dielectric is made up of carbon atoms and is influenced by a uniform electric field of $\vec{E} = 4.0\ \hat{x}$ V/m. Compute the dipole moment of a carbon atom then the polarization of the dielectric. Then find the bound charge densities.

4.2 A sphere of radius R has frozen-in polarization $\vec{P} = kr^2\ \hat{r}$ where k is a constant. Determine the bound charge densities and the electric field inside and outside the sphere. Sketch E vs. r in all regions.

4.3 An upright solid cylinder of length l and radius R has a uniform polarization parallel with its axis. find the bound charge densities.

4.4 At large distances, the voltage of a dipole in spherical coordinates can be appriximated as

$$V = \tfrac{p\cos\theta}{4\pi\epsilon_o r^2}.$$

Find the electric field for this dipole.

4.3 Linear Dielectrics

Thus far we have dealt with cases where the polarization in a dielectric could be determined from the polarizability of the atoms involved or situations where the polarization is somehow frozen-in to the material. Since electrecs are fairly rare, and it is difficult to use an atomic model to find the polarization in all cases, it becomes convenient to make a simple assumption about how the polarization in the dielectric is related to the electric field. Since earlier in this chapter it was assumed that the an atomic dipole moment was directly proportional to an external electric field it is natural then to make the approximation that the polarization in a dielectric would be directly proportional to the average electric field inside the dielectric. Though this is not always the case, it is a good approximation for small values of the electric field magnitude. This assumption then gives that

$$\vec{P} = \epsilon_o \chi_e \vec{E} \ . \tag{4.10}$$

It is a tradition to give the constant of proportionality as a product of two constants. χ_e is called the *electric susceptibility*. Dielectrics that obey Eq. (4.10) are referred to as *linear dielectrics*. The electric field in Eq. (4.10) is the net electric field as given in Eq. (4.1). It is the resultant of the external field and the polarization field inside the dielectric.

Obviously, the larger the value of electric susceptibility the greater the polarization for a given electric field. The value for the electric susceptibility is determined by the atomic environment of the dielectric material in question. Each pure material, that could serve as a dielectric, has a unique electric susceptibility value. One can

4.3. LINEAR DIELECTRICS

find these tabulated for most common dielectric materials. However, there are no less than four different ways in which the constant in Eq. (4.10) can be given and it is important to be familiar with all four.

One way is to define a new permittivity value for the dielectric, ϵ. The nice thing about this approach, is both Coulomb and Gauss's laws can be modified for use in a dielectric simply by replacing ϵ_o by ϵ. The same is true for the formula for capacitance in a parallel plate system as given in Eq. (3.27).

Since the field magnitude inside a dielectric is always less than that of the external field it must be that $\epsilon > \epsilon_o$. This leads to two other ways to denote a particular dielectric's nature. The ratio of ϵ to ϵ_o is called the *dielectric constant* K:

$$K = \frac{\epsilon}{\epsilon_o}. \tag{4.11}$$

Now K relates to χ_e as

$$K = 1 + \chi_e. \tag{4.12}$$

Finally, the dielectric constant is sometimes referred to as the *relative permittivity*, ϵ_r, where $\epsilon_r = \epsilon/\epsilon_o$. Most often, tables give values for the dielectric constant K. But, knowing how the four constants relate to one another any of the other three can be written given knowledge of one. Table 4.1 lists some values for the dielectric constant of some common pure materials.

In this text when dealing with a problem involving a dielectric where the polarization is not explicitly given, always assume that it is a linear dielectric and use Eq. (4.10) to compute $\vec{P}$.

122   CHAPTER 4. ELECTRIC FIELDS IN MATTER

Table 4.1: Dielectric constants and strengths for various pure materials at (25 °C).

Substance	Dielectric constant	Dielectric strength (V/m)
Air	1.0006	3.0×10^6
Poly(Vinyl Chloride) (PVC)	3.39	11.8×10^6
Polystyrene	2.6	24×10^6
Neoprene (rubber)	6.7	15.7×10^6
Glass	5.0	14.1×10^6
Mica	7.0	118.0×10^6
Water	80	65.0×10^6
Transformer oil	4.1	110.7×10^6

Example 4.3

A solid dielectric sphere of radius R has a point charge $+Q$ embedded at its center. Find the electric field inside and outside the sphere, the polarization and the bound charge densities. If the dielectric material is mica and $Q = 2.0$ μC what is the magnitude of the surface charge on the outer surface of the sphere?

To begin, use Gauss's law to find expressions for the electric fields inside and out. For $r < R$ Gauss's law is

$$E 4\pi r^2 = \frac{Q}{\epsilon}.$$

So, for $r < R$:

$$E = \frac{Q}{4\pi \epsilon r^2} \hat{r}.$$

and for $r > R$

$$E = \frac{Q}{4\pi \epsilon_0 r^2} \hat{r}.$$

4.3. LINEAR DIELECTRICS

We assume that it is a linear dielectric and use Eq. (4.10) to get

$$\vec{P} = \frac{\epsilon_0 \chi_e Q}{4\pi \epsilon r^2} \hat{r} = \frac{\chi_e Q}{4\pi \epsilon_r r^2} \hat{r} .$$

Use Eq. (4.5) to get the surface charge density on the outer surface. Here $\hat{n} = \hat{r}$.

$$\sigma_b = \frac{\chi_e Q}{4\pi \epsilon_r R^2} \hat{r} \cdot \hat{r} = \frac{\chi_e Q}{4\pi \epsilon_r R^2} .$$

To determine ρ_b we use Eq. (4.8) and a formula for the divergence in spherical coordinates from an Appendix to get

$$\rho_b = -\frac{1}{r^2} \frac{\partial}{\partial r} r^2 \left(\frac{\chi_e Q}{4\pi \epsilon_r r^2} \right) = 0 .$$

Now, we seek the net bound charge on the outer surface. Using Eq. (4.12) we can rewrite σ_b in terms of the dielectric constant K,

$$\sigma_b = \frac{(K-1)Q}{4\pi K R^2} .$$

So, the net charge q on the surface will be

$$q = \frac{(K-1)Q}{4\pi K R^2} (4\pi R^2) = \frac{(K-1)Q}{K} .$$

From Table 4.1 we find that for mica $K = 7.0$ so that

$$q = \frac{(7-1)(2.0 \times 10^{-6})}{7} = +1.7 \times 10^{-6} \text{ C} .$$

There is a charge of -1.7×10^{-6} C smeared about an infinitesimal surface surrounding the point charge $+Q$ at the sphere's center. Since this surface cannot be quantified, we need not consider it other than to know the bound charge is there so that for a Gaussian surface surrounding the entire sphere the net charge enclosed is $+Q$.

Note that in Table 4.1 another quantity is listed along with the dielectric constant called the *dielectric strength*. The dielectric strength has the units of an electric field. What is its purpose for being listed? Earlier we mentioned that the approximation of a linear dielectric is good for small values of the electric field magnitude. It so happens that as the electric field magnitude is increased a point is reached where the electrons begin to break free from their host atoms in the dielectric and the integrity of the material begins to fail. This phenomenon is referred to as *dielectric breakdown*. The dielectric strength is the maximum electric field magnitude that the dielectric can endure before breakdown.

PROBLEMS

4.5 Show that the capacitance, C, for a parallel plate capacitor with plate area A and inner plate distance d can be given by

$$C = \frac{AK\epsilon_o}{d},$$

where K is the dielectric constant.

4.6 A parallel plate capacitor is to be made with PVC as its dielectric. It will have plate area of 1.2 cm^2 and a plate separation distance of 1.0 mm. What is its capacitance? What is the maximum voltage that can safely be applied to the capacitor? If a voltage 25% less than the maximum safe value is applied, how much energy is stored in the capacitor?

4.7 A solid sphere of dielectric material has a hollow inner spherical cavity of radius a. At the center of the sphere is a point charge $+Q$. A cross-section of the sphere is shown

4.4. THE DISPLACEMENT FIELD

below. Find the electric field in all regions, the polarization and the bound charges.

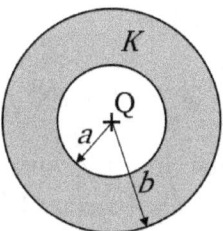

4.8 A parallel plate capacitor of plate area $A = l^2$ and inner plate separation distance d has a dielectric material of permittivity ϵ in half of the region between the plates. Derive a formula for the capacitance of this system.

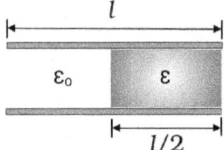

4.4 The Displacement Field

From the previous section, one can see that when a dielectric is present within an electric field there is a possibility of some bound volume charge density ρ_b being present in addition to any excess volume charge density that might have been placed within the region of interest. Therefore, we need to adjust Gauss's law to account for this additional charge. The total charge density in a region containing a dielectric, ρ, can be though of as being the sum of two charge types, the bound charge and the excess or *free charge*, ρ_f. That is,

$$\rho = \rho_b + \rho_f . \tag{4.13}$$

Recalling Gauss's law in differential form, Eq. (2.22), we have for a dielectric within a Gaussian surface in free space that

$$\vec{\nabla} \cdot \vec{E} = \frac{\rho_b + \rho_f}{\epsilon_o} . \tag{4.14}$$

Using Eq. (4.8) in this we can relate $\vec{E}$ to the polarization as

$$\vec{\nabla} \cdot \vec{E} = \frac{-\vec{\nabla} \cdot \vec{P} + \rho_f}{\epsilon_o} . \tag{4.15}$$

Rearranging and using the distributive property of the divergence, the above can be written as

$$\vec{\nabla} \cdot (\epsilon_o \vec{E} + \vec{P}) = \rho_f . \tag{4.16}$$

The sum of two vector fields yields another vector field. So, from Eq. (4.16), we define the electric *displacement field*, $\vec{D}$ as

$$\vec{D} = \epsilon_o \vec{E} + \vec{P} . \tag{4.17}$$

This new vector field is useful in that it is often more straight forward to determine its form given a known charge distribution with or without a dielectric present. Notice how in Eq. (4.17) when the polarization is zero $\vec{D} = \epsilon_o \vec{E}$ so that if $\vec{E}$ is known $\vec{D}$ is known as well and vice versa.

4.4. THE DISPLACEMENT FIELD

Using Eq. (4.14) in Eq. (4.17) one can derive a version of Gauss's law for the displacement field.

$$\vec{\nabla} \cdot \vec{D} = \rho_f . \tag{4.18}$$

On considering Eqs. (2.18) through (2.22) it is seen that the above can be written in integral form as

$$\oint_S \vec{D} \cdot d\vec{A} = Q_f , \tag{4.19}$$

where Q_f is the total free charge enclosed by the surface S.

Notice how displacement fields depends only upon the free charge density even with a dielectric present so that $\vec{D}$ is always independent of the material medium.

If the polarization is not zero we can use the approximation of a linear dielectric to relate $\vec{D}$ to $\vec{E}$. Using Eq. (4.10) to replace $\vec{P}$ in Eq. (4.17) leads to

$$\vec{D} = \epsilon_o(1 + \chi_e)\vec{E} . \tag{4.20}$$

From Eqs. (4.11) and (4.12) we conclude that

$$\epsilon = \epsilon_o(1 + \chi_e) , \tag{4.21}$$

and therefore in a dielectric medium

$$\vec{D} = \epsilon \vec{E} . \tag{4.22}$$

The displacement field is especially useful for situations involving a static electric field and a dielectric medium. For example, suppose one wanted to find the voltage, V, in all regions, inside and outside, of a material polarized by an electric field. First you might want to know the electric field in all regions and then use Eq. (3.13) to compute V. To use Gauss's law to get $\vec{E}$, both the free and bound charges are required. Of course, with the polarization known, it is an easy thing to find the bound charge. However, if it is a linear dielectric, knowledge of the electric field is required to get $\vec{P}$! This can all be circumvented by first writing the expression for $\vec{D}$ in all regions as this vector field is independent of bound charge. Then, it is a straight forward thing to use Eq. (4.22) to get $\vec{E}$.

Example 4.4

Consider the hollow sphere with the interior point charge $+Q$ as shown in Problem 4.7. Find the voltage within the region $a < r < b$.

We'll get $\vec{E}$ then use Eq. (3.13) to compute $V(r)$. It is a straight forward thing to use Gauss's law to find $\vec{D}$ in all regions and then write $\vec{E}$. Though $\vec{D}$ and $\vec{E}$, within the region $r < a$, are not required for this problem, it is easy to use Eq. (4.19) to arrive at

$$\vec{D} = \frac{Q}{4\pi r^2} \quad \text{for} \quad r < a.$$

In the region $a < r < b$ the free charge enclosed is still just $+Q$ so that

$$\vec{D} = \frac{Q}{4\pi r^2} \quad \text{for} \quad a < r < b.$$

Obviously, this is the same result for $\vec{D}$ outside the sphere. Now for $r < a$ there is free space so we use $\vec{D} = \epsilon_o \vec{E}$ and thus

$$\vec{E} = \frac{Q}{4\pi \epsilon_o r^2} \quad \text{for} \quad r < a.$$

4.4. THE DISPLACEMENT FIELD

In the region $a < r < b$ we use Eq. (4.22) to get

$$\vec{E} = \frac{Q}{4\pi\epsilon r^2} \quad \text{for} \quad a < r < b .$$

Now, to get V we apply Eq. (3.13) in the two regions, $a < r < b$ and $r > a$:

$$V = -\int_R^r \frac{Q}{4\pi\epsilon r^2} dr - \int_\infty^R \frac{Q}{4\pi\epsilon_o r^2} dr ,$$

After integrating an rearranging one gets

$$V = \frac{Q}{4\pi} \left[\frac{1}{\epsilon_o R} + \frac{1}{\epsilon}\left(\frac{1}{r} - \frac{1}{R}\right) \right] .$$

Let's compute the bound charge densities while we're at it. Using $\vec{E}$ within the polarized region, $a < r < b$, in Eq. (4.10) $\vec{P}$ is determined as

$$\vec{P} = \frac{\epsilon_o \chi_o Q}{4\pi\epsilon r^2} \hat{r} .$$

There will be bound surface charge at $r = a$ and $r = b$. For $r = b$, $\hat{n}$ points in the positive direction so that by Eq. (4.5),

$$\sigma_{bb} = \frac{\epsilon_o \chi_o Q}{4\pi\epsilon b^2} \hat{r} \cdot \hat{r} = \frac{\epsilon_o \chi_o Q}{4\pi\epsilon b^2} .$$

At the surface $r = a$, $\hat{n}$ is directed inwards so that

$$\sigma_{ba} = \frac{\epsilon_o \chi_o Q}{4\pi\epsilon a^2} \hat{r} \cdot -\hat{r} = -\frac{\epsilon_o \chi_o Q}{4\pi\epsilon a^2} .$$

Using Eq. (4.8), and the divergence in spherical coordinates, we find for the bound volume charge density

$$\rho_b = -\frac{1}{r^2} \frac{\partial}{\partial r} r^2 \left(\frac{\epsilon_o \chi_o Q}{4\pi\epsilon r^2}\right) = 0 .$$

4.5 Electrostatic Boundary Conditions

Interestingly, there was a discontinuity in the magnitude of the electric field vector at the interface between the dielectric surface and air in Example 4.4. This discontinuity exist for all electric field vectors that pass between two non-conducting materials. This phenomenon yields an interesting consequence for the optical properties of materials so it is profitable to consider the situation further here.

This effect can be described analytically by considering the results from Example 4.4 for the electric field inside and outside the dielectric sphere. Rather than consider a dielectric and air we generalize the problem by considering two non-conducting regions of permittivity ϵ_1 and ϵ_2. Referring back to Example 4.4, let region one be $a < r \leq b$ and region two be $r > b$. From Example 4.4 we can then write the electric field in each region at the interface where $r = b$ as

$$\vec{E}_1 = \frac{Q}{4\pi\epsilon_1 b^2} \hat{r} , \qquad (4.23)$$

and

$$\vec{E}_2 = \frac{Q}{4\pi\epsilon_2 b^2} \hat{r} , \qquad (4.24)$$

Obviously, $|\vec{E}_1| \neq |\vec{E}_2|$. However, letting $|\vec{E}_1| = E_1$ and $|\vec{E}_2| = E_2$ it must be that at the interface between the two regions $\epsilon_1 E_1 = \epsilon_2 E_2$. Of course, both $\vec{E}_1$ and $\vec{E}_2$ are normal to the interface so we will denote this fact by letting the magnitudes of Eqs. (4.23) and (4.24) be E_{1n} and E_{2n} respectively. Therefore we must have that

$$\epsilon_1 E_{1n} = \epsilon_2 E_{2n} . \qquad (4.25)$$

4.5. ELECTROSTATIC BOUNDARY CONDITIONS

We started this section by considering the electric field vector crossing a spherical surface but that need not be the case. The results found here will be valid whatever the geometry of the surface between the two nonconducting media so that we will henceforth assume that the interface between the two regions is flat planar in the area of the vector's location.

Now, let's generalize the problem so that the electric field vector has a normal and tangential component at the interface. This situation is depicted in Figure 4.2. There

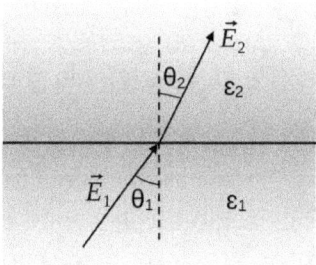

Figure 4.2: An electric field vector at the interface between two nonconducting regions.

is a discontinuity in the normal component of the field vector at the interface which is described by Eq. (4.25). It can be shown that the component of the vector tangent (parallel) to the interface is not discontinuous but rather remains constant as the field vector crosses the boundary. We denote the tangential components of $\vec{E}_1$ and $\vec{E}_2$ as E_{1t} and E_{2t} respectively. Therefore

$$E_{1t} = E_{2t} . \qquad (4.26)$$

Referring to Figure 4.2, the permittivities in the two regions can be related to the angles shown in the figure. From Figure 4.2 we have that

$$\tan\theta_1 = \frac{E_{1t}}{E_{1n}}, \qquad (4.27)$$

and

$$\tan\theta_2 = \frac{E_{2t}}{E_{2n}}. \qquad (4.28)$$

Using Eq. (4.26) in the above leads to

$$E_{1n}\tan\theta_1 = E_{2n}\tan\theta_2. \qquad (4.29)$$

Now, using Eq. (4.25) in this yields the desired relation

$$\epsilon_2\tan\theta_1 = \epsilon_1\tan\theta_2. \qquad (4.30)$$

Suppose that one region was a solid conductor holding excess charge. From previous discussions we know that this charge would reside on the surface of the conductor and the electric field inside the conductor must be zero. Therefore, the field outside the conductor is just the field due to a flat plane of surface charge density σ, so that the tangential component of $\vec{E}$ is zero on both sides of the interface. Thus the electric field in the non-conducting region, bounding the charged conductor, is always normal to the interface.

PROBLEMS

4.9 An electrec in the shape of a solid sphere of radius R has a frozen-in polarization given by $\vec{P} = P_o r \, \hat{r}$ where P_o is a constant. Determine all bound charge densities, $\vec{D}$ and $\vec{E}$ in all regions.

4.10 For the coaxial cable considered in Problem 2.26, find $\vec{D}$ in all regions.

4.11 Determine $\vec{D}$ between the plates of the parallel plate capacitor considered in Example 2.4.

4.12 Suppose you have two nonconducting regions as depicted in Figure 4.2. Let region one be polystyrene and region two be glass. If $\theta_1 = 20°$ what is θ_2?

4.13 Electrostatic boundary conditions can sometimes be used to find the electric field inside a dielectric if the electric field is known outside. Suppose a thin slab of a dielectric material is placed upright within a uniform horizontally directed electric field $\vec{E}$. Find an expression for the electric field inside the dielectric in terms of the outside field. (Ignore fringing effects at the top and bottom edges of the slab.)

4.6 Energy in Dielectrics

The results considered in the previous chapter for energy stored within electric fields can easily be modified to deal with cases where a dielectric is present. Often, such as for the parallel plate capacitor, one need only replace ϵ_o, in the expression being used to compute the energy, by the permittivity value, ϵ, for the dielectric in question. For example, the energy stored in a dielectric filled capacitor would be simply given by Eq. (3.31) with capacitance replaced by the expression verified in Problem 4.5.

But what about a more general case where we seek the energy stored in the electric field in the presence of a dielectric? In this situation, Eq. (3.38) gets the job done. However, there are two ways this can be accomplished. First, considering an electric field in free space due to the free and bound charges, which is labeled as $\vec{E}_{fb}$, we simply reuse Eq. (3.38) as it stands,

$$U = \frac{\epsilon_o}{2} \int_\tau E_{fb}^2 d\tau . \quad (4.31)$$

Another approach is to deal with an electric field within the dielectric due to the free charge alone, denoted as $\vec{E}_f$ which gives,

$$U = \frac{\epsilon}{2} \int_\tau E_f^2 d\tau . \quad (4.32)$$

Now, since in a linear medium $\vec{D} = \epsilon \vec{E}$, the above can be written as

$$U = \frac{1}{2} \int_\tau \vec{D} \cdot \vec{E}_f d\tau . \quad (4.33)$$

If there is no free charge, as can be the case with an electrec, then one must use Eq. (4.31) to compute the stored energy.

4.6. ENERGY IN DIELECTRICS

Example 4.5

Consider the energy stored in a parallel plate capacitor charged by some voltage V with a linear dielectric between the plates as shown in Figure 4.1. Find this energy using Eq. (4.31) and Eq. (4.33) and show that the energy found via Eq. (4.33) is larger than that of Eq. (4.31) by a factor of K. Give an explanation for this discrepancy.

In Eq. (4.31) we use only the assemble free and bound charge in free space so that

$$U_1 = \frac{\epsilon_o}{2} \int \left(\frac{\sigma+\sigma_b}{\epsilon_o}\right)^2 d\tau \ .$$

When using Eq. (4.33) we need just the field of the free charge but within the medium of the dielectric so, $D = \sigma$ and $E = \sigma/\epsilon$. Using this in Eq. (4.33) along with a little factoring we get

$$U_2 = \frac{\epsilon}{2} \int \left(\frac{\sigma}{\epsilon}\right)^2 d\tau \ .$$

It is left as a problem to show that for a linear dielectric

$$\frac{\sigma+\sigma_b}{\epsilon_o} = \frac{\sigma}{\epsilon} \ .$$

The integrands in the above expressions are constant and can be factored out. Letting $\tau = \int d\tau$ leads to

$$U_1 = \frac{\epsilon_o}{2} \left(\frac{\sigma}{\epsilon}\right)^2 \tau \ .$$

and

$$U_2 = \frac{\epsilon}{2} \left(\frac{\sigma}{\epsilon}\right)^2 \tau \ .$$

Multiplying U_2 by ϵ_o/ϵ_o leads to

$$U_2 = KU_1 \ .$$

We find that $U_2 > U_1$ as Eq. (4.33) accounts for the additional work required to polarize the atoms in the dielectric. Eq. (4.31) does not. Therefore, it is typically the best choice to use Eq. (4.33) to compute the energy stored in a polarized linear dielectric.

4.7 The Clausius-Mossotti Equation

As a final topic for this chapter we will discuss the derivation of the *Clausius-Mossotti equation*. This relationship connects the atomic polarizability of a dielectric material to its dielectric constant. This in turn makes the determination of the atomic polarizability for a given material a simple matter of measuring the capacitance of a capacitor while the substance serves as a dielectric between the capacitor plates.

The result we want is obtained by considering the relationship between the electric susceptibility, χ_e, for a linear dielectric and the atomic polarizability, α. Recall that for a linear dielectric $\vec{P} = \epsilon_o \chi_e \vec{E}$ where $\vec{E}$ is the macroscopic net field within the dielectric. However, the dipole moment for an atom in the dielectric is

$$\vec{p} = \alpha \vec{E}_{ext} . \qquad (4.34)$$

Here $\vec{E}_{ext}$ is the microscopic field at the atom, as we ignore the field due to the dipole itself. We seek to relate $\vec{P}$ to $\vec{p}$ and thus learn about the connection between χ_e and α. To do so break the electric field into two parts, one the field of the dipole $\vec{E}_d$ and the the other, the remaining portion of the field external to the dipole, $\vec{E}_{ext}$. So, the total macroscopic field within the dielectric is

$$\vec{E} = \vec{E}_d + \vec{E}_{ext} . \qquad (4.35)$$

4.7. THE CLAUSIUS-MOSSOTTI EQUATION

Let us assume that $\vec{E}_d$ can give by the field of a lone electric dipole. From Problem 4.4 this is

$$\vec{E}_d = \frac{p}{4\pi\epsilon_o r^3}\left(2\cos\theta\,\hat{r} + \sin\theta\,\hat{\theta}\right). \tag{4.36}$$

Notice that the above is a function of the angle θ. This complicates the derivation so that we employ a standard simplifying approximation for Eq. (4.36) given by

$$\vec{E}_d = -\frac{\vec{p}}{4\pi\epsilon_o R^3}, \tag{4.37}$$

where R is the radius of a sphere that surrounds the dipole which can be assumed to be the atomic radius. The negative sign has been added since the field of the dipole will be directed opposite that of $\vec{E}_{ext}$. Now, $\vec{p}$ reacts to the external field, not the field of itself so that $\vec{p} = \alpha \vec{E}_{ext}$. Using this in Eq. (4.37), and inserting into Eq. (4.35) leads to

$$\vec{E} = \left(1 - \frac{\alpha}{4\pi\epsilon_o R^3}\right)\vec{E}_{ext}. \tag{4.38}$$

Using Eq. (4.34) in Eq. (4.4) leads to

$$\vec{P} = N\alpha\vec{E}_{ext}, \tag{4.39}$$

where N is the density of atoms in the dielectric. Using $N = 3/(4\pi R^3)$ to replace R and then solving Eq. (4.38) for $\vec{E}_{ext}$, and inserting this into Eq. (4.39), yields

$$\vec{P} = \frac{N\alpha}{(1 - N\alpha/3\epsilon_o)}. \tag{4.40}$$

On comparing this result with Eq. (4.10) we conclude that for a linear dielectric

$$\chi_e = \frac{N\alpha}{\epsilon_o(1 - N\alpha/3\epsilon_o)}. \tag{4.41}$$

Eq. (4.41) can be solved for α. It is tradition to write this result in terms of the dielectric constant, K, thus arriving at the Clausius-Mossotti Equation:

$$\alpha = \frac{3\epsilon_o}{N}\left(\frac{K-1}{K+2}\right). \qquad (4.42)$$

PROBLEMS

4.14 Use the Clausius-Mossotti equation to show that the polarizability of a dielectric can be given by

$$\alpha = \frac{3\epsilon_o}{N}\left(\frac{1}{1+\frac{3C}{\Delta C}}\right),$$

where C is the free space capacitance of a parallel plate capacitor and ΔC is its change in capacitance upon insertion of the dielectric. What is the polarizability as $\Delta C \to \infty$?

4.15 Derive Eq. (4.42) from Eq. (4.41).

4.16 A spherical conductor of radius r_1 which holds charge $+Q$ is embedded in the center of a larger sphere of radius R made of a linear dielectric material. Find the stored energy.

4.17 Calculate the energy stored in a sphere of radius R of uniform dielectric material with a frozen-in polarization $\vec{P} = P_o r\,\hat{r}$ where P_o is a constant.

4.18 Referring back to Example 4.5, show that for a parallel plate capacitor with a linear dielectric

$$\frac{\sigma}{\epsilon} = \frac{\sigma + \sigma_b}{\epsilon_o},$$

for $\sigma_b < 0$.

4.7. THE CLAUSIUS-MOSSOTTI EQUATION

4.19 For an electrostatic field $\vec{\nabla} \times \vec{E} = 0$. Is $\vec{\nabla} \times \vec{D} = 0$? If not, under what circumstances is the curl of $\vec{D}$ not zero?

4.20 A parallel plate capacitor has two linear dielectric materials between the plates as shown in the figure below. The plates hold surface charge density σ. Find $\vec{E}$ and $\vec{D}$ in both interior regions. Find all the bound charges and the total energy stored. Hint: Can this system be treated like capacitors in parallel or series?

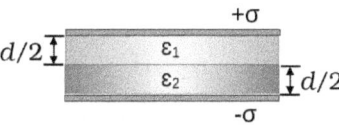

4.21 A solid conducting sphere of radius r_1 holds charge $+Q$. It is isolated by a dielectric from an outer solid conducting spherical shell which holds charge $-Q$. See the figure below.

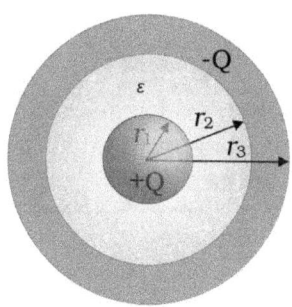

a) Find $\vec{E}$ and $\vec{D}$ in all regions. Sketch E vs. r and D vs. r for all regions.
b) Find the voltage at the center of the inner sphere and the capacitance of this assembly.
c) Find the polarization and the bound charge densities.

4.22 Refer to the coaxial cable discussed in Problem 2.26. Assume that the region $a < r < b$ is now filled with a dielectric material of dielectric constant K while the inner solid conductor holds charge $+Q$ and the outer thin shell holds $-Q$. Find $\vec{D}$ and $\vec{E}$ in all regions. Also find the polarization and the bound charge densities.

4.23 An upright cylinder of radius R has a frozen-in polarization $\vec{P} = P_o r^3 \hat{r}$ where P_o is a constant. Determine the bound charge densities.

4.24 Consider the interface between two different non-conducting materials. There is a free surface charge density σ at the interface. Use Gauss's law for the displacement field, Eq. (4.19), to show that there is a discontinuity in the normal component of the displacement field at the boundary given by

$$D_{1n} - D_{2n} = \sigma \ .$$

Therefore, if there is no free charge at the interface

$$D_{1n} = D_{2n} \ .$$

4.25 A conducting sphere which holds charge $+Q$ is immersed halfway into a nonconducting liquid. Find $\vec{E}$ and $\vec{D}$ in all regions and the free surface charge density.

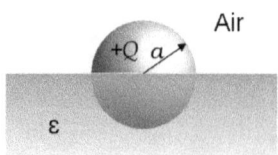

4.7. THE CLAUSIUS-MOSSOTTI EQUATION

4.26 A dielectric slab is made of PVC. Determine values for its electric susceptibility and permittivity. This dielectric slab is now used to fill the space between the plates of a capacitor which is given a voltage of 500 V. The distance between the plates is 0.52 cm. Find the polarization and the bound surface charge density. What is the maximum safe voltage for this capacitor?

4.27 Two parallel conducting plates, infinite in the y and z directions are separated by a distance d. A constant voltage V_o is applied to the plates. There is a dielectric between the plates which has a permittivity that varies with position and is given by

$$\epsilon(x) = \frac{4\epsilon_o}{\left(\frac{x}{d}\right)^2 + 2} \ .$$

Find $\vec{E}$ and $\vec{D}$ within the dielectric, the polarization and the bound volume charge density.

4.28 A solid conducting sphere of radius R which holds charge $+Q$ is embedded within an infinite dielectric medium of permittivity ϵ. Compute the total stored energy.

4.29 Consider the situation discussed in Problem 4.28 only now the dielectric medium has a permittivity given by

$$\epsilon(r) = \epsilon_o \left(1 + \frac{R}{r}\right)^3 \ ,$$

where $r = 0$ is at the center of the conducting sphere. Find the total stored energy in this case.

142   CHAPTER 4.   ELECTRIC FIELDS IN MATTER

4.30 Show that the change in stored energy, ΔU, in a capacitor upon insertion of a dielectric slab can be given by

$$\Delta U = U_o \left(1 - \tfrac{1}{K}\right),$$

where U_o is the initial stored energy with free space between the plates.

4.31 The potential energy, U, of a dipole moment, $\vec{p}$, in an electric field, $\vec{E}$, can be given by $U = -\vec{p} \cdot \vec{E}$. Suppose the electric field is the time dependent field $\vec{E}(t) = E_o \sin(\omega t) \, \hat{x}$, where t is time and ω an angular frequency. If $\vec{p}$ points in the same direction as $\vec{E}$ show that

$$U = -\alpha E_o^2 \sin^2(\omega t).$$

At what times t will $U = 0$? Where does the potential energy go at these times?

4.32 A uniform electric field $\vec{E} = E_x \, \hat{x} + E_y \, \hat{y}$ interacts with the electric dipole $\vec{p} = p_x \, \hat{x} + p_y \, \hat{y}$. If θ is the angle between the two vectors, show that the Schwarz inequality leads to

$$(E_x^2 + E_y^2)(p_x^2 + p_y^2) \cos^2 \theta \le (E_x^2 + E_y^2)(p_x^2 + p_y^2).$$

Show that the potential energy, $-\vec{p} \cdot \vec{E}$, is a minimum when $\theta = 0$ and thus the Schwarz inequality for $\vec{p} \cdot \vec{E}$ becomes an equality when the potential is at its minimum.

4.33 The voltage of a static dipole is given by

$$V = \tfrac{p \cos \theta}{4\pi \epsilon_o r^2}.$$

Find the electric field for this voltage. What is the electric field at large distances from the dipole?

Chapter 5

Magnetostatics

5.1 Current

The compass has been in use as a navigational tool since at least the 11th century and perhaps even earlier. However, as far as is known to the author, the presentation of a useful quantitative description of the phenomenon which produces the behavior of the compass, is a relatively recent occurrence. It was in the early 19th century that the Danish physicist Hans Christian Oersted (1777- 1851) discovered that a current in a wire would cause a compass needle to deflect.

A compass needle aligning along the north-south direction on Earth and a compass needle being deflected by a current in a wire are both different examples of a phenomenon we refer to as *magnetism*. Initially, many physicists thought that magnetism was a distinct phenomenon not necessarily related to static charges, Coulomb's law or the electric field. We now understand that magnetism and electricity are one and the same phenomenon only seen from different perspectives. However, it is still useful to study magnetism as a separate subject and this is the approach taken in this and following chapters.

As with static charges and the electric field in previous chapters, the properties of magnetism can be described using a vector field. Whereas static charges are the source for all static electric fields, charges moving at constant speed are the source for all static magnetic fields. Therefore, it is essential that we study the motion of charge particles before considering the nature of the static magnetic vector field.

The flow of charged particles, for example, electrons, through space is referred to as *current*. To quantify this effect consider a group of electrons flowing at constant velocity, $\vec{v}_d$ through a uniform electric field as shown in Figure 5.1. We can assume

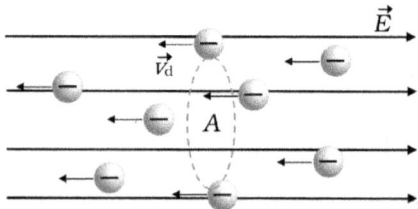

Figure 5.1: Electrons flow to the left under the influence of a uniform electric field.

that the medium is air or some other gas for the moment. Also, we will assume that any acceleration the electrons experience can be neglected and the justification for this simplification will be given later.

For such as situation, the *current*, I, is defined as the charge, q, passing through some area, A, per unit time, t. That is

$$I = \frac{dq}{dt} . \tag{5.1}$$

The SI unit for current will be the Coulomb per second better known as the Amp (A).

It is useful to have a vector field that describes such a current. If the density of the *charge carriers*, in this case

5.2. DRUDE THEORY OF CONDUCTIVITY

electrons, is given by n, then a current density, $\vec{J}$, can be defined as

$$\vec{J} = -nq\vec{v}_d , \qquad (5.2)$$

where q is the charge of the electron. The velocity $\vec{v}_d$ is more formally known as the *drift velocity* or sometimes the *drift speed*. The current, I, is then simply the flux of the current density, $\vec{J}$, through some surface S,

$$I = \int_S \vec{J} \cdot d\vec{A} . \qquad (5.3)$$

Example 5.1

A uniform current density in free space is given by $\vec{J} = (4.0\,\hat{x} + 3.0\,\hat{y})\,10^{-3}\,\mathrm{A/m^2}$. Find the current through a square surface of side length 0.25 m that lies in the xz plane.

Let $d\vec{A}$ point in the positive direction then $d\vec{A} = dxdz\,\hat{y}$. Using Eq. (5.3) leads to

$$I = \int_0^{0.25} \int_0^{0.25} (4.0\,\hat{x} + 3.0\,\hat{y})\,10^{-3} \cdot \hat{y}\,dxdz = $$
$$\int_0^{0.25} \int_0^{0.25} 3.0 \times 10^{-3}\,dxdz = (0.25)^2(3.0 \times 10^{-3}) = 0.89\,\mathrm{mA} .$$

5.2 Drude Theory of Conductivity

Since electrons readily flow though conductors, a phenomenon referred to as *electrical conductivity*, it is simple to create the conditions for a steady current within a solid metal. The required environment is that of a constant electric field within the sample. In this section we will discuss the behavior of such a current and briefly review a classical theory given to describe the fundamental physical mechanisms behind the availability of free carriers within a metal and their subsequent behavior under the influence of an electric field.

In 1900, around three years after the discovery of the electron by J. J. Thompson (1856-1940), Paul Drude (1863-1906) proposed a theory that described the process of electrical conduction in metals, the so-called *Drude theory of conductivity*. We break the model down into three main postulates here.

1. The electrons in the metal are either core electrons or valence/conduction electrons. Core electrons are tightly bound to their host atoms while valence electrons are free to move about the entire sample.

2. The valence electrons behave as an electron gas and as such can be dealt with using ideas from the kinetic theory of gases. These gaseous electrons are held within a container of size given by the bulk dimensions of the metal sample.

3. The valence electrons undergo scattering events that maintain thermal equilibrium and thus some mean thermal velocity v_m with a mean time between scattering events given by τ.

Let us examine each of these three postulates in greater detail. In this discussion we will only consider pure, elemental metals. Postulate one tells us that we can esti-

5.2. DRUDE THEORY OF CONDUCTIVITY

mate the density of conduction electrons in a pure sample by considering the valence of the metal from the periodic table. This suggests that the conduction electrons per atom will be one for the alkali metals, Li, Na and K and also for the metals Cu, Ag and Au. Also, from the table, we find that Al has valence three which suggests three free conduction electrons per atom. However, for the transition metals, more care must be taken as many of the valence electrons are paired via the Pauli principle and as such do not participate in the electron gas. For example, though iron has valence five, a count of 2 free electrons per atom is more in agreement with the Drude model description of its observed electrical conductivity.

Denoting the number of conduction electrons donated per atom as Z, it is then a straight forward thing to compute the carrier density, n, for various pure metals using the following formula,

$$n = \frac{ZN_A\rho_m}{W_a} . \qquad (5.4)$$

Here N_A is Avogadro's number, ρ_m the mass density and W_a the atomic weight.

The second Drude postulate tells us that the free electrons are held within a potential well the length and width of which is equal to the size of the sample. At room temperature the average kinetic energy of the conduction electrons is not sufficient to escape this potential well. However, at higher temperatures such electrons can be *boiled off*, a process not of interest to us in this text.

Finally, perhaps the most intriguing of the three postulates, tells us that the conduction electrons are maintained at thermal equilibrium due to some scattering event. The Drude model does not give details on the nature of this scattering event, one simply accepts that such activity occurs. Obviously, it is likely that the conduction electrons have collisions with the atomic cores and other electrons but these details need not be considered in the application of this model. What is important is that there will be some mean thermal speed for the electrons, v_m, and a mean time between collisions, τ.

When one applies a uniform electric field of magnitude E to the system of conduction electrons in a metal by Netwon's law they will experience an acceleration magnitude given by qE/m where q is electron charge and m the electron mass. Assuming that the mean time between scattering events remains constant the final speed at time τ induced on the electrons by the field is, v_d, and is given by

$$v_d = \frac{qE\tau}{m} . \tag{5.5}$$

This is simply our drift speed, or drift velocity, mentioned earlier in this chapter. In most metals, $v_m >> v_d$ so that this drift speed amounts to only a small change in the speed of the conduction electrons in the metal.

On comparing Eq. (5.5) with Eq. (5.2) we conclude that

$$\vec{J} = -\frac{nq^2\tau}{m}\vec{E} . \tag{5.6}$$

The scalar constants in Eq. (5.6) can be set equal to a value called the *conductivity*, σ. That is,

$$\sigma = \frac{nq^2\tau}{m} , \tag{5.7}$$

where one is careful not to confuse the use of σ here with that of a surface charge density in earlier chapters. Each

5.2. DRUDE THEORY OF CONDUCTIVITY

Table 5.1: Electrical resistivities for various pure materials at (25 °C). Taken from, D.R. Lide, CRC Handbook of Chemistry and Physics, 88th ed., CRC Press, 2008.

Substance	Resistivity ρ_r (Ωm)
Copper	1.7×10^{-8}
Aluminum	3.7×10^{-8}
Zinc	6.0×10^{-8}
Iron	9.7×10^{-8}
Nickel	10.0×10^{-8}
Titanium	43.0×10^{-8}
Stainless steel	72.0×10^{-8}
Carbon	1.4×10^{-5}
Germanium	46.0×10^{-2}
Silicon	3.0×10^{-2}

pure metal has a unique conductivity value. Furthermore, it is independent of the size of the sample. Often, this constant is given as a *resistivity*, ρ_r where $\rho_r = 1/\sigma$. Resistivity values for various pure materials are listed in Table 5.1.

Eq. (5.6) gives the current density for electron flow. It is a tradition that current density is taken to be due to the flow of positive charge. This perspective introduces no difficulties so long as one remembers that the flow of this fictitious positive charge is opposite that of the electrons. With this convention in mind, the negative sign in Eq. (5.6) is removed leaving a famous relation known as *Ohm's law*:

$$\vec{J} = \sigma \vec{E} . \tag{5.8}$$

5.3 Resistance

One can see from Eq. (5.8) the importance of a conductivity or resistivity value as this quantity determines the magnitude of current density within a sample given a known electric field.

Often, it is useful to have a measure of resistance to flow that is dependent on the size of the sample in question. With metals one often deals with a cylindrical sample of constant circular cross-section, a so-called wire. If the cross-sectional area of such a wire is A then a differential *resistance*, dR, can be given as

$$dR = \frac{\rho_r dl}{A} . \tag{5.9}$$

If ρ_r is constant, then for a wire of length l the above integrates to yield a value for the resistance R of a wire of length l and cross sectional area A.

$$R = \frac{\rho_r l}{A} . \tag{5.10}$$

Example 5.1

Consider a solid cone shaped metal sample of uniform resistivity ρ_r as shown below. The upper end is circular with radius b while the lower circular end has radius a. The height of the cone is l. Derive a formula for the resistance of this sample. What does your result give for the case where $a = b$? What happens when $a \to 0$?

We use the differential form of Eq. (5.9) where the area is a function of the axial coordinate which in this case is z. Therefore we have

5.3. RESISTANCE

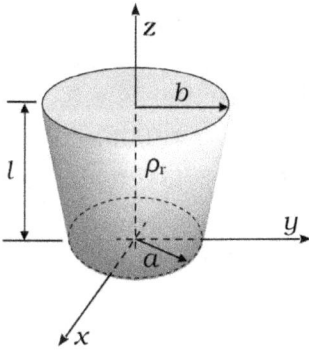

$$R = \int_0^l \frac{\rho_r}{A(z)} dz .$$

The radius of the cone as a function of z can be given by the line

$$y = \left(\frac{b-a}{l}\right) z + a .$$

Putting it all together we have

$$R = \int_0^l \frac{\rho_r}{\pi \left[\left(\frac{b-a}{l}\right)z+a\right]^2} dz .$$

Using a variable substitution the above integrates to

$$\begin{aligned} R &= -\frac{\rho_r l}{\pi(b-a)} \left[\frac{1}{\left(\frac{b-a}{l}\right)z+a} \right]\Bigg|_0^l \\ &= \frac{\rho_r l}{\pi(a-b)} \left(\frac{a-b}{ab}\right) = \frac{\rho_r l}{\pi ab} . \end{aligned}$$

When $a = b$ we get, $R = \rho_r l/(\pi a^2)$ as expected from Eq. (5.10). When $a \to 0$, $R \to \infty$.

In North America, wire cross-section diameters are often given in gauge sizes of the *American Wire Gauge*

Scale (AWG) also sometimes referred to as the Brown and Sharpe scale. For example, a 12 gauge wire in this scale has a diameter of approximately 0.2 cm. This size would be listed as 12 AWG. A listing of AWG values, along with corresponding diameters given in centimeters, is shown in Table 5.2. The gauge sizes 0 AWG, 00 AWG and 000 AWG are read as single, double and triple naught, respectively.

Table 5.2: American wire gauge scale (AWG).

Gauge	diameter (cm)	Gauge	diameter (cm)
0000	1.168	7	0.3665
000	1.040	8	0.3264
00	0.9266	9	0.2906
0	0.8252	10	0.2588
1	0.7348	11	0.2305
2	0.6543	12	0.2053
3	0.5827	13	0.1828
4	0.5189	14	0.1628
5	0.4620	15	0.1450
6	0.4115	16	0.1291

Occasionally, the electron drift velocity can be more useful than a current density especially when dealing with semiconductor materials. For our purposes, semiconductors are the pure elements Si, Ge and Carbon which have resistivity values between those of a conductor and an insulator. Therefore, Eq. (5.5), which gives the drift velocity as being directly proportional to the magnitude of the applied field, is most useful here. Combining the

5.3. RESISTANCE

Table 5.3: electron mobilities for various pure semiconductor materials at (25 °C). Taken from, D.R. Lide, CRC Handbook of Chemistry and Physics, 88th ed., CRC Press, 2008.

Substance	Mobility cm^2/(V s)
Si	1900
Ge	3800
Carbon	1800

constants, this can be written as

$$\vec{v}_d = \mu \vec{E} . \tag{5.11}$$

The constant of proportionality, μ, is referred to as the electron *mobility*. Using Eqs. (5.5) and (5.7), and the fact that $\rho_r = 1/\sigma$, leads to

$$\mu = \frac{1}{\rho_r n q} . \tag{5.12}$$

Mobility values for the above mentioned semiconductor materials are listed in Table 5.3.

PROBLEMS

5.1 Compute the electron carrier density for copper. Then, using data from Table 5.1 find the mean time between electron scattering events.

5.2 Using data from Tables 5.1 and 5.3, estimate the electron carrier density for silicon. Compare this value to that of copper found in Problem 5.1.

5.3 In a 10 AWG copper wire the current density is 1.3 A/m^2.

a) What is the current in the wire?
b) What is the electric field in the wire?

5.4 A 12 AWG aluminum wire carries a current of 1.2 A.

a) What is the current density in the wire?
b) What is the electron mobility in the wire?
c) What is the electron drift speed?

5.5 A wire, of circular cross-sectional area A, has a resistivity ρ_r given by $\rho_r(x) = \rho_o e^{kx}$ where ρ_o and k are constants. x is the axial distance along the wire. What will be the resistance for a length l of this wire?

The form of Ohm's law given above could be called the field form of Ohm's law. Another commonly used version is where $\vec{J}$ is replaced by I. This can be called the *circuit* version of Ohm's law as it it primarily used to deal with calculations involving electric circuits. Eq. (5.8) is easily converted to this form by letting $\vec{E}$ be a uniform field as between two capacitor plates, only here in the wire, which is possible as charges are not at rest. We consider some length l of a wire of constant cross-sectional area A. Since the field is uniform the magnitude of $\vec{E}$ can be given by the voltage, V, over the length l. Then, Eq. (5.8) can be rewritten as

$$\frac{I}{A} = \sigma \frac{V}{l}. \qquad (5.13)$$

5.3. RESISTANCE

Using the fact that $\rho_r = 1/\sigma$ and Eq. (5.10) the above becomes the circuit form of Ohm's law:

$$I = RV .\tag{5.14}$$

Since we are dealing with steady currents, it must be that energy is being dissipated to the surroundings as the current flows through a resistance. This is much like an object in mechanics moving at constant velocity due to friction. Heat is dissipated to the surroundings in both cases. It is useful to then have an expression for the power, P, or energy dissipated by the current per unit time passing through a resistive medium.

Recalling that voltage is an energy, U, per charge, q, we can use Eq. (5.1) to rewrite Eq. (5.14) as

$$\frac{U}{q} = R\frac{dq}{dt} .\tag{5.15}$$

Power is given by $P = dU/dt$ so that multiplying both sides of Eq. (5.15) by q and differentiating with respect to t leads to

$$P = R\left[q\frac{d^2q}{dt^2} + \left(\frac{dq}{dt}\right)^2\right] .\tag{5.16}$$

As we are dealing with steady currents, the second derivative in the above must vanish leaving the *Joule heating law*:

$$P = I^2R .\tag{5.17}$$

Example 5.2

Consider a cone shaped sample of solid carbon as depicted in Example 5.1 where $a = 1.00$ mm, $b = 0.50$ cm and $l = 2.20$ cm. A steady current density of 12.0 $\hat{z}$ A/m² enters the bottom face at $z = 0$. Assume that current can flow freely into and out of the cone ends as if it is a component in a circuit. What is the current in the sample? What is the current density at the upper face? What power is dissipated in the solid?

First compute I through the bottom face,

$$I = (12.0)\pi(1.00 \times 10^{-3})^2 = 37.7 \ \mu A.$$

The current I is conserved through the entire sample so at the top face $\vec{J}$ will be

$$\vec{J} = \frac{37.7 \times 10^{-6}}{\pi(0.50 \times 10^{-2})^2} \hat{z} = 0.48 \ \hat{z} \ A/m^2.$$

Getting the resistivity of carbon from Table 5.1 and using the formula for resistance derived in Example 5.1, the resistance of the sample is computed as

$$R = \frac{(1.4 \times 10^{-5})(2.20 \times 10^{-2})}{\pi(1.00 \times 10^{-3})(0.50 \times 10^{-2})} = 19.6 \ m\Omega.$$

Finally, we use the Joule heating law to compute the power.

$$P = (37.7 \times 10^{-6})^2(19.6 \times 10^{-3}) = 2.79 \times 10^{-11} \ W.$$

To conclude this brief section on current, we will derive an important formula that relates the current density to the time varying volume charge density that generates the current. Recall from Eq. (5.3) that the current I

5.3. RESISTANCE

is simply the flux of $\vec{J}$ through some surface S. By the fundamental theorem for divergences we can then write

$$\int_S \vec{J} \cdot d\vec{A} = \int_\tau \vec{\nabla} \cdot \vec{J} \, d\tau . \tag{5.18}$$

Since $I = dq/dt$ we can state that

$$\int_\tau \vec{\nabla} \cdot \vec{J} \, d\tau = \frac{dq}{dt} . \tag{5.19}$$

The charge, q, entering or exiting the volume, τ, can be given as the integral over the volume charge density ρ so that Eq. (5.19) becomes

$$\int_\tau \vec{\nabla} \cdot \vec{J} \, d\tau = -\frac{\partial}{\partial t} \int_\tau \rho \, d\tau . \tag{5.20}$$

Here partial differentiation is used as ρ can be a function of position and time. We want a positive divergence when charge is leaving the volume therefore a minus sign was introduced on the right side of Eq. (5.20).

The arguments of the integrals in Eq. (5.20) must be equivalent so that

$$\vec{\nabla} \cdot \vec{J} = -\frac{\partial \rho}{\partial t} . \tag{5.21}$$

Eq. (5.21) is a statement of the conservation of charge. Equations of this type, where the spatial change in some quantity is connected with its variance over time, are often referred to as *continuity equations*. Such relationships are very powerful and occur in many different areas of physics including fluid mechanics, heat and mass transfer and quantum mechanics.

158 CHAPTER 5. MAGNETOSTATICS

PROBLEMS

5.6 A solid cylinder is capped with a solid hemispherical end with a flat top the plane of which is parallel with its bottom face. Both sections are made of the same material having resistivity ρ_r, (see the figure below). A current is to flow along the direction of the z axis. Derive a formula for the resistance of this sample.

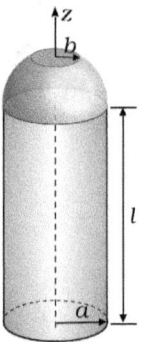

5.7 Suppose there is a time dependent charge flow within a wire such that $q(t) = q_o \sin(\omega t)$ where q_o and ω are constants. Use Eq. (5.16) to show that the power in this case can be given by

$$P = R q_o^2 \omega^2 \left(1 - 2\sin^2(\omega t)\right) .$$

Use the mean value theorem from calculus to compute the average power. Comment on this result.

5.8 Compute the Joule heating loss (also referred to as i-squared r loss) in a 35.0 km long 000 AWG aluminum cable that has voltage of 20 kV relative to ground and provides current for a load of 2.0 kΩ. How much energy is lost from this line in a year?

5.9 A cylindrical sample of pure silicon of length l has a free electron density n_o. This natural carrier density is sometimes referred to as the *intrinsic* carrier density. Through a process called *doping*, whereby impurities are added to silicon, the carrier density of silicon can be increased. Suppose the cylindrical sample of silicon is doped so that the carrier density, n, in the sample of silicon varies with axial location x along the cylinder as

$$\frac{dn}{dx} = n_\beta \beta e^{-\beta x},$$

where, n_β and β are constants and at the sample end, $n(0) = n_o$. Find a formula for the total carrier density in the sample. Use the resistivity and mobility of pure silicon to estimate n_o. Let $n_\beta = 1.0 \times 10^{20}$ m^{-3} and $\beta = 200.0$ m^{-1}. Find the total carrier density in a 3.0 mm long cylinder of the doped silicon sample. Compute the new electron mobility for this doped sample.

5.10 A current density is given by $\vec{J} = 4x\,\hat{x} + 5e^{-2y}\,\hat{y}$ A/m^2. Determine the charge density that generates such a current.

5.4 The Biot-Savart Law

Having now, studied the essentials of steady current flow, we can come to a description of the phenomenon of magnetism through knowledge of the behavior and shape of the magnetic field. As stated earlier in this chapter, the source of all statics magnetic fields are steady currents.

Steady currents are often referred to as direct currents, or DC currents.

One can get a qualitative description of the shape of the magnetic field by using a right hand rule (RHR). Place the thumb of the right hand in the direction of positive current flow. Then, the fingers of the right hand will curl in the direction of the magnetic field lines. This activity is depicted in Figure 5.2.

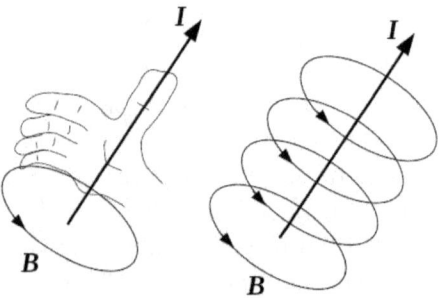

Figure 5.2: Using the RHR to help visualize the magnetic fields lines due to steady line current I. On the right, the complete field lines are drawn.

Notice how the magnetic field lines form closed circular loops around the line current. We call such a field a *circumferential field*. It is an interesting property of static magnetic field lines that they always form closed loops. As with electric field lines, at every point along a field line there is a field vector tangent to the line thus forming a vector field. For the magnetic field, we label this vector field as $\vec{B}$.

As Coulomb's law was the postulate used to describe

5.4. THE BIOT-SAVART LAW

the static electric field, there is another postulate known to be effective in yielding a quantitative description of a static magnetic field. This is the *Biot-Savart law.* Unlike Coulomb's law, there is no sort of magnetic charge involved here. Rather, the Biot-Savart law gives us a way to write the magnetic vector field due to a steady line current I.

Consider a line current that need not necessarily be straight. Let $d\hat{l}$ be an infinitesimal line vector along the path of the current. The Biot-Savart law gives the infinitesimal for the magnetic field vector $d\vec{B}$ at some point, P, away from the current due to the differential current at $d\hat{l}$. This situation is sketched in Fig. 5.3. $\vec{r}$ is a vector

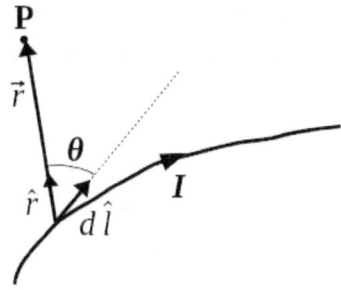

Figure 5.3: Details required for implementing the Biot-Savart law.

from the point of $d\hat{l}$ to the point of interest P. $\hat{r}$ is a unit vector in the direction of $\vec{r}$. Let r be the magnitude of $\vec{r}$. The Biot-Savart law states that

$$d\vec{B} = \frac{\mu_o I}{4\pi} \frac{d\hat{l} \times \hat{r}}{r^2} . \tag{5.22}$$

Here the constant μ_o is called the *permeability of free space*. The SI unit for the magnetic field magnitude, B, is the Tesla (T). With this in mind, the value for μ_o in SI units will be $4\pi \times 10^{-7}$ T·m/A.

With θ being the angle between $d\hat{l}$ and $d\hat{r}$, we can use a property of the cross product to write for the magnitude of $d\vec{B}$,

$$dB = \frac{\mu_o I dl \sin\theta}{4\pi r^2} . \tag{5.23}$$

Example 5.3

Use the Biot-Savart law to find the magnitude of the magnetic field due to a circular DC current loop of radius R at any point along an axis through its center and perpendicular to the plane of the loop.

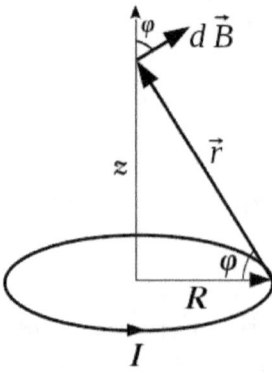

Figure 5.4: Details required for implementing the Biot-Savart law for a current loop.

5.4. THE BIOT-SAVART LAW

Here at all points around the loop $d\hat{l}$ is perpendicular to $\vec{r}$ so that $|d\hat{l} \times \hat{r}|$ reduces to dl. The distance r will be given by $\sqrt{z^2 + R^2}$. Using polar coordinates in the plane of the loop we write dl as $Rd\theta$. Using Eq. (5.22) we have that

$$\int_0^B dB' = \frac{\mu_0 IR}{4\pi(z^2+R^2)} \int_0^{2\pi} d\theta \;.$$

Unfortunately, the above expression will give an incorrect answer. Since $\vec{B}$ is a vector, the sum around the loop will incorrectly add into the result all of the components of $\vec{B}$ that are perpendicular to the z axis. Looking at the figure, we see that due to symmetry, the perpendicular components of $\vec{B}$ cancel. We therefore, adjust the above expression so that it sums only the components of $\vec{B}$ that are parallel to the z axis. We extract these desired components by multiplying the above by $\cos\phi$. From the figure we see that

$$\cos\phi = \frac{R}{\sqrt{z^2+R^2}} \;.$$

Putting it all together we get

$$\int_0^B dB' = \frac{\mu_0 IR^2}{4\pi(z^2+R^2)^{3/2}} \int_0^{2\pi} d\theta \;.$$

So that

$$B = \frac{\mu_0 IR^2}{2(z^2+R^2)^{3/2}} \;.$$

In the plane of the loop, where $z = 0$, the field has its maximum value along the z axis:

$$B = \frac{\mu_0 I}{2R} \;.$$

PROBLEMS

5.11 Use the Biot-Savart law to show that the magnetic field magnitude a perpendicular distance r above an infinite straight DC line current, I, is given by

$$B = \frac{\mu_0 I}{2\pi r}.$$

5.12 A long thin strip of width w carries a DC current I. Assume that the current is evenly distributed across the strip. Show that at a point a perpendicular distance y directly above its center the field magnitude can be given by

$$B = \frac{\mu_0 I}{\pi w} \arctan\left(\frac{w}{2y}\right).$$

5.13 Use the Biot-Savart law to find the magnitude of the magnetic field at the center of a square current loop of width w.

5.5 Magnetic Forces

One of the distinct features of a magnetic field is the manner in which it interacts with a free charge. As with an electric field, there is a field force acting upon the free charge but, of an entirely different nature.

It is useful to begin by considering uniform magnetic vector fields. For example, consider a uniform magnetic field directed into the page as depicted in Figure 5.5. Now, suppose a particle of charge $+q$ is traveling into this field with constant velocity directed from left to right as shown below in the same figure. As with the test charge used to study the Coulomb force, we ignore the field of

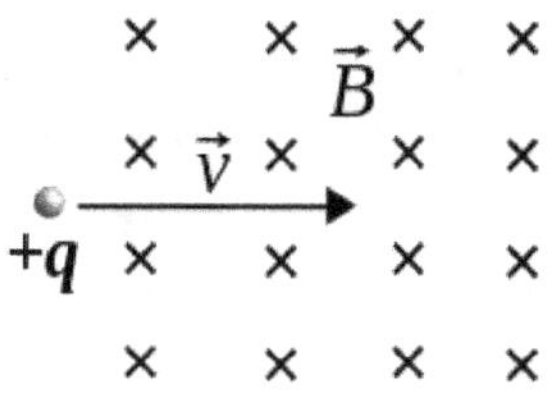

5.5. MAGNETIC FORCES

```
    ×    ×  →×   ×
            B
    × →v ×      ×   ×
  •  ─────────────▶
 +q  ×    ×      ×   ×

    ×    ×      ×   ×
```

Figure 5.5: A charged particle moving with velocity $\vec{v}$ enters a uniform magnetic field directed into the page.

the moving free charge. The particle will experience a force due to the magnetic field, which we label as $\vec{F}_B$, which is given by

$$\vec{F}_B = q(\vec{v} \times \vec{B}) \ . \qquad (5.24)$$

From Eq. (5.24), and using the right hand rule for cross products, we can determine the direction of the resulting force. For the situation depicted in Figure 5.5, the magnetic force on the particle is directed straight up in the plane of the page.

Eq. (5.24) reveals the two features that make the magnetic force remarkably different from the Coulomb force. First, notice how the formula involves the particle's velocity. Therefore, if the particle's velocity is zero, with respect to the magnetic field, then the force vanishes. Also, since the formula involves a cross product there is a geometric element involved as well. Recall, that when two vectors lie along the same line, (that is, they are coterminal), their cross product yields zero. So when the particle travels in the direction of a magnetic field line, the magnetic force vanishes.

Recall that the magnitude of a cross product can be written in terms of the angle θ between the two vectors and their magnitudes. Therefore, from Eq. (5.24) we have that $F_B = qvB\sin\theta$. In Figure 5.5, $\theta = 90°$. In this special case, where $\vec{v}$ is perpendicular to $\vec{B}$, we have that

$$F_B = qvB . \tag{5.25}$$

For the situation depicted in Figure 5.5, so long as the charged particle remains in the magnetic field $\vec{F}_B$ will be perpendicular to $\vec{v}$. Thus $\vec{F}_B$ plays the role of a centripetal force while $\vec{v}$ is the tangential velocity. If the magnitude of $\vec{v}$ remains constant, the particle will undergo uniform circular motion. This situation is depicted below in Figure 5.6. The period for this rotation, T, can

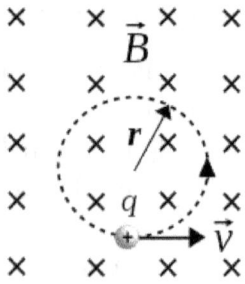

Figure 5.6: A charged particle with constant speed, $|\vec{v}|$, undergoes uniform circular motion within a uniform magnetic field directed into the page.

be related to the magnitude of the magnetic field. Recall that for uniform circular motion the period is given by

$$T = \frac{2\pi r}{v} . \tag{5.26}$$

5.5. MAGNETIC FORCES

Here v is the magnitude of the particle's tangential velocity. We can get the magnitude of the uniform field involved by considering the relation from mechanics for centripetal force magnitude and then equate this to the magnetic force magnitude. This give

$$m\frac{v^2}{r} = qvB , \qquad (5.27)$$

Solving for v in Eq. (5.27), and using this in Eq. (5.26, yields

$$T = \frac{2\pi m}{qB} . \qquad (5.28)$$

Since the frequency, f, in Hertz, is the reciprocal of the period, we then have that the frequency of the orbit is

$$f = \frac{qB}{2\pi m} . \qquad (5.29)$$

This result is often referred to as the *cyclotron frequency*.

If an electrostatic field is also present, and assuming that the principle of superposition holds, then the two forces can be added to yield the resultant force, $\vec{F}$, acting on a free charge, $+q$. This resultant force is referred to as the *Lorentz force*:

$$\vec{F} = q\vec{E} + q(\vec{v} \times \vec{B}) . \qquad (5.30)$$

Example 5.4

Show that the net work done on a moving charge within a static magnetic field is zero.

From the work-energy theorem of mechanics we have that the net work, W, done on a particle of mass, m, is equal to the particle's change in kinetic energy. Letting the particle's initial speed be v_i and the final speed be v_f the work-energy theorem is

$$W = \tfrac{1}{2}mv_f^2 - \tfrac{1}{2}mv_i^2 \ .$$

Once the charged particle enters the magnetic field its speed can be given by solving for v in Eq. (5.27). That is

$$v = \tfrac{qBr}{m} \ .$$

Since, q, B and r all remain constant it must be the the speed of the charged particle remains unchanged during the orbit so that $v_f = v_i$ for any selected time interval. Therefore, by the work-energy theorem, the net work done on the particle by the magnetic field is zero.

So, unlike the electric field, the magnetic field alters only the direction of the particle, not its speed.

PROBLEMS

5.14 A particle holding charge $+2.0$ μC with velocity $\vec{v} = (\hat{x} + 2.0\,\hat{y} - 3.0\,\hat{z})\,10^3$ m/s, enters a region with an electric field $\vec{E} = 4.0\,\hat{x} - 2.0\,\hat{y}$ V/m and a magnetic field $\vec{B} = (0.001)\,\hat{x} + 2.0\,\hat{y}$ T. What is the Lorentz force acting on the particle?

5.15 Consider a particle of charge $+q$ moving with velocity $\vec{v} = v_o\,\hat{y}$ that enters a region where there is a uniform magnetic field directed into the page, that is $\vec{B} = -B_o\,\hat{x}$ and a uniform

5.5. MAGNETIC FORCES

electric field directed left to right on the page, $\vec{E} = E_o \, \hat{y}$. (consider a three dimensional Cartesian system as shown in Figure 1.1). Describe the path the particle takes for the three cases: 1) $qE < qvB$, 2) $qE = qvB$ and 3) $qE > qvB$.

5.16 Consider a situation, within two dimensional Cartesian free space, where a particle of charge $+q$ and mass m moving with velocity $\vec{v} = v_x \, \hat{x} + v_y \, \hat{y}$ is within a uniform electric field $\vec{E} = E_x \, \hat{x} + E_y \, \hat{y}$ and a uniform magnetic field $\vec{B} = B_x \, \hat{x} + B_y \, \hat{y}$ while also experiencing a force due to gravity given by $-mg \, \hat{y}$. Show that the net force acting on the particle is zero when

$$qE_x = 0$$
$$qE_y = -mg \text{ and}$$
$$v_x B_y = v_y B_x \, .$$

Even though the ultimate source of a static magnetic field is a steady current, there are moving charges within a current so that when an external line current is placed within a magnetic field there will be a magnetic force acting on the line current.

It is useful to have a mathematical description for this phenomenon. Such is easily arrived at by using Eq. (5.24). Consider some length of the line current l. Let Q be the charge in length l. If the speed of the current is v then we can write that $I = \frac{Qv}{l}$. Letting $\vec{l}$ be a vector with magnitude l in the direction of $\vec{v}$ we have that

$$\vec{l} = \frac{Q\vec{v}}{I} \, . \tag{5.31}$$

Now, consider two parallel line currents, I_1 and I_2, separated by some distance d. The force on I_2 due to the field $\vec{B}_1$ of I_1 will be $\vec{F}_1$. Likewise the field, $\vec{B}_2$, due to I_2 exerts a force on I_1, $\vec{F}_2$. On comparing Eq. (5.31) with Eq. (5.24) we have that the force $\vec{F}_1$ can be written as

$$\vec{F}_1 = I_1 \vec{l} \times \vec{B}_2 . \qquad (5.32)$$

Likewise for $\vec{F}_2$

$$\vec{F}_2 = I_2 \vec{l} \times \vec{B}_1 . \qquad (5.33)$$

Of course, in the case where $\vec{I}$ is perpendicular to $\vec{B}$ the magnitudes for both of the above reduce to $F = lIB$.

More generally, the force, $\vec{F}_l$, on a line current due to the magnetic field $\vec{B}$ per unit length can be given by

$$\vec{F}_l = I\hat{l} \times \vec{B} . \qquad (5.34)$$

The above expressions are valid for a DC current in a straight wire and a uniform magnetic field. For a DC current in a curved wire and a magnetic field that varies with position we can generalize the above with

$$d\vec{F} = I d\hat{l} \times \vec{B} . \qquad (5.35)$$

where $d\hat{l}$ is the infinitesimal line vector.

Example 5.5

A rigid line current of 3.0 A in the shape shown in the figure below runs through a uniform magnetic field of magnitude $B = 1.2$ T. Here $l = 0.5$ m and $r = 0.25$ m. Determine the net force acting on the wire.

The magnitude of the force on one of the straight sections is simply IlB. By the right hand rule for cross products we

5.5. MAGNETIC FORCES

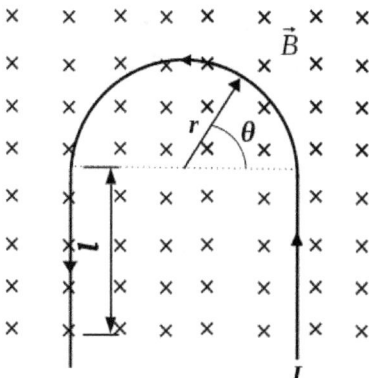

Figure 5.7: A line current in a uniform magnetic field. the line current is composed of two parts straight line and one part that is semicircle.

find that the force on the right straight section will be equal and opposite to the force on the left hand straight section. Therefore, the forces on the straight sections cancel and can be ignored.

From the orientation of the line current and Eq. (5.35) we see that the angle between $\vec{dl}$ and $\vec{B}$ will always be 90° so that the differential force magnitude is

$$dF = IdsB .$$

where ds is the infinitesimal differential arc distance along the circular path. This magnitude will have a component in the horizontal and vertical direction in the figure shown above. Note how the horizontal components on the right will cancel those on the left. Therefore, we only need sum the vertical component of the force around the semi-circular path. The figure below shows how the force at a point along the path relates to the angle θ.

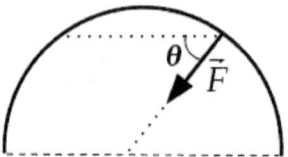

Figure 5.8: The vertical component of the magnetic force on the wire can be related to the angle θ.

Therefore, the vertical component of the force will be given by $F\sin\theta$. Since r is constant, a formula from geometry gives $ds = rd\theta$. Now, we sum the contributions up from $\theta = 0$ to $\theta = \pi$.

$$F = IBr \int_0^\pi \sin\theta d\theta = -IBr\cos\theta|_0^\pi = 2BIr \ .$$

Inserting our values for the field magnitude, current and radius leads to

$$F = 2(1.2)(3.0)0.25 = 1.8 \text{ N} \ .$$

5.6 The Velocity Selector

A device that utilizes the forces exerted on charged particles, by electric and magnetic fields, is the *velocity selector*. Given a collection of charged particles in the gaseous state, typically ions of atoms or molecules, this device can be used to select only those particles with a translational speed within a certain range. For purposes of discussion, it will be assumed that the selector is perfect

5.6. THE VELOCITY SELECTOR

and that only charged particles of one particular speed are selected.

The charged particles usually under consideration are the ions of well known molecules. These can be cations, those having a net positive charge, or anions, those having a net negative charge. You may recall from a previous study of the distribution of molecular speeds in a gas that at any instant each molecule in the ensemble will be undergoing translation with a unique speed taken from a range of speeds. This dispersion of speeds can be model by the Maxwell-Boltzmann distribution of speeds.

The ions used in the velocity selector are typically created through some sort of combustion process and we will assume that afterwards they are left in the gaseous state with a distribution of speeds. So how does the velocity selector act to select only one of the speeds in the distribution? By select here it is meant that the device allows only particles of a particular speed to pass and be emitted from the selector. This process is accomplished by the simultaneous use of an electric and magnetic field. The setup is depicted below in Fig. 15.9.

From the nature of the electric field, we know that the force magnitude, F_E, it exerts on the cation will be directed downward. From the right hand rule for cross products, it is seen that the force on the particle due to the magnetic field, of magnitude F_B, is directed upwards. Therefore, the particle will not be deflected when these two forces are equal and opposite. This implies that

$$qE = qvB . \tag{5.36}$$

Solving this for v gives

$$v = \frac{E}{B} . \tag{5.37}$$

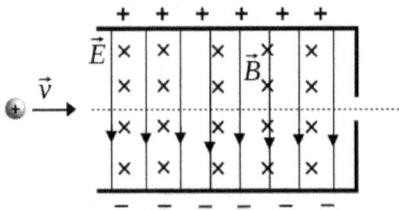

Figure 5.9: A cation traveling with speed v enters a velocity selector. A uniform magnetic field points into the page while a uniform electric field points downwards. The particle will travel through the center of the exit passage only if the magnetic force on the particle equals the electric field force.

Therefore, by selecting the appropriate value for the ratio E/B, one can use the velocity selector to emit only charged particles of a particular speed. This device is a necessary part of a mass spectrometer, an instrument used to measure the mass of various molecules. The spectrometer uses a velocity selector to set the speed of particles which then enter a region with a second uniform magnetic field. In this second field, the particles are allowed to curve in a circular path of radius r. The particles then strike a detector which can be used to determine the value for r. With r and the speed v known, the mass of the charged particle can be determined.

5.6. THE VELOCITY SELECTOR

Example 5.6

The mass spectrometer. The charged plates separated by a distance of 0.10 m in a velocity selector are given a voltage of 1.0 kV. +1 cations leave the velocity selector with a speed of 1.0×10^4 m/s they then enter an external uniform magnetic field of magnitude B_{ex}. This situation is shown schematically in the figure below.

a) What is the magnitude of the magnetic field in the velocity selector?

b) Derive an expression for the mass m of the cation in terms of the relevant parameters such as the radius of the particle orbit and the magnitude of the external field outside the velocity selector?

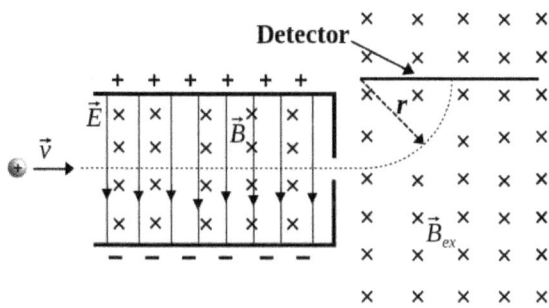

Figure 5.10: Diagram showing the essential features of a mass spectrometer.

a) Here we use $E = V/d$ to compute the electric field magnitude in the selector

$$E = \frac{V}{d} = \frac{1000}{0.10} = 100.0 \text{ V/m} .$$

Now, solving Eq. (5.37) for B we get

$$B = \frac{E}{v} = \frac{100}{1.0\times 10^4} = 0.01 \text{ T}.$$

b) Once the cation leaves the selector it will experience a force due to the external field B_{ex} with magnitude given by qvB_{ex}. Here q is the $+1$ charge, that is, the charge of one proton. This force is a centripetal force so using the know expression for the magnitude of the centripetal acceleration we can write the following equality of force magnitudes:

$$qvB_{ex} = m\frac{v^2}{r}.$$

Solving this for m we get the desired formula

$$m = \frac{qrB_{ex}}{v}.$$

PROBLEMS

5.17 A circular current loop of radius R carries the current I and lies perpendicular to a uniform magnetic field $\vec{B}$. (See the figure below). Show that the net magnetic force acting on the loop is zero. (Hint: Consider an example problem in this chapter.)

5.6. THE VELOCITY SELECTOR

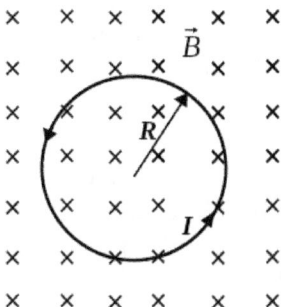

15.18 A line current of 2.2 A, in the shape of a parabola, passes through a 120.0 cm section of a uniform magnetic field as shown below. Here the magnitude of the magnetic

field is 1.2 T. The line current path can be described by $f(x) = x^2$. Compute the net magnetic force magnitude acting on the wire. (Hint: Here $d\hat{l}$ is always perpendicular to $\vec{B}$ so that $d\hat{l} \times \vec{B} = B dl$. Also, use the formula from calculus for the arc length along a curve described by $f(x)$, that is: $\int_a^b \sqrt{1 + f'(x)^2}\, dx$.)

15.19 A line current of 2.0 A goes in the direction $\vec{l} = 2.0\,\hat{x} - 4.0\,\hat{y}$ m. It is surrounded by a magnetic field given by $\vec{B} = 0.023\,\hat{x} + 0.2\,\hat{z}$ T. What is the magnetic force per unit length acting on the current?

15.20 For a velocity selector, the magnetic field magnitude is 1.5 T between charged plates a distance 3.5 cm apart. If you desire to select a particle speed of 2.5×10^3 m/s what must the voltage be on the plates?

15.21 A velocity selector emits +1 cations with a speed of 1.0×10^4 m/s into a mass spectrometer which has a uniform magnetic field of magnitude 2.0 T. The spectrometer detects two orbital radii, one of 0.26 mm the other 0.416 mm. What are the masses of the two particles detected? Since these particles are cations of atoms, molecules or fragments of molecules there mass should be some multiple of the proton's mass. Report their mass as a multiple of the proton's mass. (The mass of the electrons can be ignored.)

5.7 The Magnetic Moment

Thus far we have considered the circumferential (solenoidal) magnetic field due to a line current and situations where the field can be assumed to be uniform, (without specifying details for the current that created the uniform field). In this section, we will study a magnetic field with yet a different shape which originates due to a current flowing in a circular path. This particular magnetic field can be used to describe a variety of magnetic phenomena. These include the behavior of the compass, magnets and

5.7. THE MAGNETIC MOMENT

the magnetic behavior of atoms, molecules and atomic particles.

Suppose that a DC current is flowing through a circular loop as depicted in the figure below. Using the right hand rule for magnetic fields, you can envision the appearance of the magnetic field lines surrounding the loop. It is similar to the shape of a pumpkin.

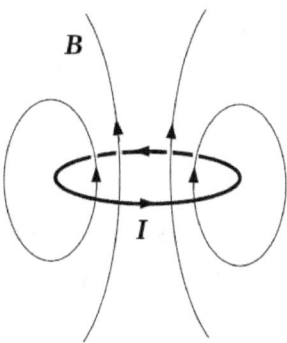

Figure 5.11: Selected magnetic field lines, and sections of field lines, around a current loop.

When dealing with circular current loops, it is useful to define a quantity called the *magnetic dipole moment* $\vec{\mu}$, or simply the magnetic moment. If the area inscribed by the circular current loop is A then the magnetic moment is defined as

$$\vec{\mu} = IA\hat{n} , \qquad (5.38)$$

where $\hat{n}$ is a unit vector that is perpendicular to the plane of the area A. To determine the direction of $\hat{n}$ and thus $\vec{\mu}$, cup the fingers of your right hand in the direction of the current flow in the loop. Then, your thumb points in the direction of $\hat{n}$. So for the current loop depicted in Figure 5.11, the magnetic dipole moment points upward. The SI unit for $\vec{\mu}$ would be the amp-meter squared (Am^2).

An interesting feature of magnetic field lines is that they always form closed loops. (To save space, we've elected to show only sections of some of the closed loops in Figure 5.11.) This is distinctly different from electric field lines. Recall, that one isolated charge can produce an electric field. The electric field lines are directed radially away from the lone charge in all directions never forming closed loops. We refer to an isolated static charge as an electric monopole. So it must be that no known magnetic monopoles exist. This is another way of saying that magnetic field lines always form closed loops.

As with the electric field, the word dipole implies there must be two entities involved. For the electric dipole there are two charge types, in the magnetic case we declare there are two pole types. The convention is to label these poles as north (N) and south (S). One can think of poles as a sort of magnetic charge. As with charge and the electric field, there is a convention for the direction of a magnetic field with respect to the pole type. Magnetic field lines run from north to south poles. These poles are related to the magnetic moment vector in that the magnetic moment points from south to north magnetic pole. The magnetic dipole vector for the current loop of Figure 5.11 is added to this figure along with the denotations for north (N) and south (S) pole and shown in Figure 5.12. Notice how the magnetic moment points from the south to north magnetic pole.

Recall, there was a force between charged particles,

5.7. THE MAGNETIC MOMENT 181

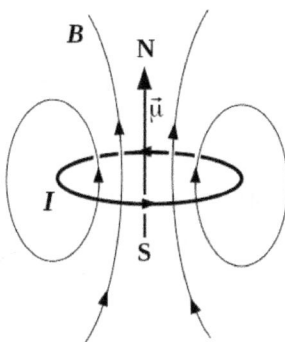

Figure 5.12: Selected magnetic field lines around a current loop with the magnetic dipole vector and poles denoted.

the Coulomb force. The Coulomb force can be attractive or repulsive depending upon the charge sign. Also recall from Eq. (5.24) the nature of the magnetic force–it involves a current and a field. This force law can be generalized such that qualitative predictions can be made regarding the direction of the force given knowledge of the magnetic poles involved. For example, the magnetic force between current loops can be attractive or repulsive depending upon the pole types involved. The simple rule is **like poles repel, unlike poles attract.**

It is often simpler to deal with a magnetic field in terms of its magnetic moment vector alone. The magnetic moment vector for the field given in Figure 5.11 holds much of the information necessary for determining how this field interacts with another magnetic field. This simplifies the analysis of the interaction between the current loop and neighboring magnetic fields. The magnetic moment will also be a convenient model for demonstrating the nature of magnetic materials later in this chapter.

For example, suppose we want to consider the behavior of the current loop shown in Figure 5.11, once placed within an external uniform magnetic field. Let the external field be held fixed while the current loop is free to rotate. What will happen to the current loop? To answer this question we would only need to consider the behavior of the loop's magnetic dipole moment vector within the external magnetic field. This is depicted graphically in Figure 5.13.

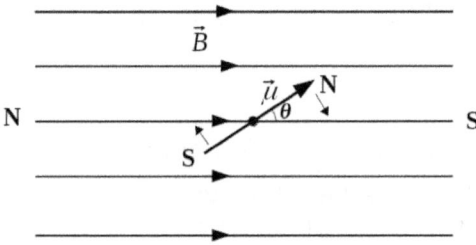

Figure 5.13: Magnetic dipole in a uniform magnetic field. Initially the dipole is inclined by the angle of θ from a field line. Since unlike poles attract the dipole will experience a torque and rotate to become aligned with a field line.

Let the magnetic moment vector be initially canted at an angle θ from a magnetic field line. Since unlike poles attract, the dipole will experience a torque and rotate about its center until it aligns with a field line. It can be shown that the torque $\vec{\tau}$ can be given by

$$\vec{\tau} = \vec{\mu} \times \vec{B} . \tag{5.39}$$

We can now comment on the behavior of the compass.

5.7. THE MAGNETIC MOMENT

The compass needle acts like a magnetic dipole and aligns with an (approximately uniform) magnetic field line that lies along the north-south direction near the surface of the Earth. This statement leads to two questions. One, how is it that a solid compass needle can behave as a current loop dipole? Secondly, why is there a magnetic field above the surface of the Earth?

We will venture an answer to the first of these two questions in this text. To answer this question we need to consider a simple model of the hydrogen atom. In this model, the nucleus is a proton and the lone electron orbits this proton in a circular path. Since the proton has many times the mass of the electron we'll assume that it is at rest with respect to the electron. This simple model is sketched in Fig. 5.14.

Since an electron is a charged particle, when it orbits the proton it creates a current moving along a circular path. The electron has a negative charge so this current can be thought of as a positive current moving in the direction opposite that of the electron.

This atomic circular current flow creates a magnetic dipole vector as with the current loop studied previously. If this is a reasonable model for the hydrogen atom then this atom should posses a non-zero magnetic moment. This is indeed observed to be the case. In fact, all atoms and molecules are observed to have a magnetic dipole moment due to electron motion about the nuclei. This dipole is often called the orbital magnetic dipole moment.

Additionally, experiment has shown that the electron itself has a non-zero magnetic dipole moment, that is, a sort of intrinsic magnetic dipole. Since the electron has charge, this can be thought of as being due to the

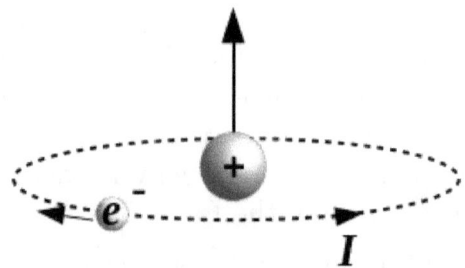

Figure 5.14: An electron orbits a proton in a model of the hydrogen atom. This electron current can be thought of a current of positive charge I in the opposite direction. This creates a current loop and thus a magnetic dipole moment.

electron spinning about an internal axis. We therefore refer to this magnetic dipole as the electron spin magnetic dipole or simply, electron spin. So, ignoring any intrinsic magnetic moment of the nucleus, the overall magnetic dipole moment of an atom or molecule can be given as a combination of the electron spin magnetic moments and the electron orbital magnetic moments.

5.8 Ampère's Law

The Biot-Savart law gives us a prescription for determining the magnetic field vectors at all locations around a line current. Therefore, we should be able to examine the divergence and curl of such a vector field. It turns out that both of these operations lead to results that play a major role in the theory of magnetic fields.

5.8. AMPÈRE'S LAW

Recall that one of the peculiar things about a magnetic field is that the field lines always form closed loops. This then implies that the flux of a magnetic field $\vec{B}$ through a closed surface S must be zero. That is

$$\oint_S \vec{B} \cdot d\vec{A} = 0 \ . \tag{5.40}$$

By the fundamental theorem for divergences this can be written as

$$\oint_S \vec{B} \cdot d\vec{A} = \int_\tau \vec{\nabla} \cdot \vec{B} \, d\tau = 0 \ . \tag{5.41}$$

The above is true when the argument of the second integral is zero so that the divergence for any magnetic field must be zero:

$$\vec{\nabla} \cdot \vec{B} = 0 \ . \tag{5.42}$$

The above result is often referred to as *Gauss's law for magnetism*. This result is a direct result of the fact that there are no known magnetic monopoles, as discussed previously.

We can examine the curl of $\vec{B}$ by considering the magnetic field due to a line current verified in Problem 5.11. In cylindrical coordinates, this is

$$\vec{B} = \frac{\mu_o I}{2\pi r} \hat{\phi} \ . \tag{5.43}$$

The nature of the curl of such a field is revealed if we consider the line integral of $\vec{B}$ around a closed path which encloses the current. That is,

$$\oint \vec{B} \cdot d\vec{l} \ . \tag{5.44}$$

It is convenient to consider a circular path of radius r where the line current passes through the center of the circular surface enclosed by the path. Along this path, $\vec{dl} = rd\phi\, \hat{\phi}$. Using such a trajectory, along with Eq. (5.43) in Eq. (5.44), leads to

$$\oint \frac{\mu_o I}{2\pi r} \hat{\phi} \cdot rd\phi\, \hat{\phi} = \mu_o I \ . \tag{5.45}$$

The result above can be shown to be true even when the current is not centered in the circular surface. The integral may be more difficult to evaluate but one can be assured that the result will be $\mu_o I$. Assuming the principle of superposition holds for the magnetic field, the same could be done for any number of line currents that pass through our circular loop. We label the sum of these enclosed currents as I_{enc}. Then in general the following is true for a closed path around I_{enc}

$$\oint \vec{B} \cdot \vec{dl} = \mu_o I_{enc} \ . \tag{5.46}$$

This result is known as the integral form of *Ampère's law*. This form of Ampère's law can be used, in certain cases, to find an expression for the magnetic field due to a current more quickly, and with less computation, than when applying the Biot-Savart law.

One can arrive at the differential form for Ampère's law, and also finally learn about the curl of $\vec{B}$, by using the fundamental theorem for curls on Eq. (5.46). That is,

$$\oint \vec{B} \cdot \vec{dl} = \oint_S \vec{\nabla} \times \vec{B} \cdot \vec{dA} \ . \tag{5.47}$$

5.8. AMPÈRE'S LAW

This allows us to write that

$$\oint_S \vec{\nabla} \times \vec{B} \cdot d\vec{A} = \mu_o I_{enc} . \quad (5.48)$$

Using Eq. (5.3) in this leads to

$$\oint_S \vec{\nabla} \times \vec{B} \cdot d\vec{A} = \mu_o \oint_S \vec{J}_{enc} \cdot d\vec{A} . \quad (5.49)$$

On comparing the integrands in Eq. (5.49) we are lead to the final result.

$$\vec{\nabla} \times \vec{B} = \mu_o \vec{J}_{enc} . \quad (5.50)$$

Example 5.7

As depicted in Figure 5.15, a straight cylindrical conductor of circular cross-section (a wire) of radius R carries current I which is uniformly distributed throughout the conductor. (Recall from our study of electric charge that excess charge in a conductor resides on the surface. Here however, there is no excess charge and the current is composed of all free carriers in the material. These can be thought of as uniformly distributed throughout the solid). Use Ampère's law to find an expression for the magnetic field inside and outside the conductor.

In the region $r < R$ we use a circular loop concentric with the center of the conductor and perpendicular to the axial. The current enclosed will be some fraction of the total current I depending upon the value for r. Since we assume that the current is spread uniformly throughout the conductor, a ratio of areas will give us the required fraction, that is

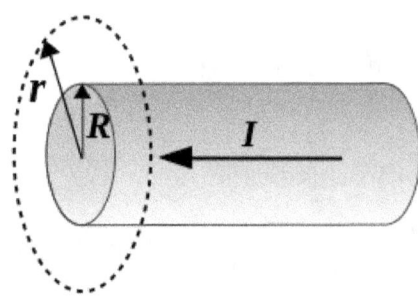

Figure 5.15: A DC current flowing through a solid cylindrical conductor of radius R.

$$I_{enc} = I \frac{\pi r^2}{\pi R^2}.$$

Along the enclosing path, $\vec{B}$ will always be in the same direction as $\vec{dl}$ and, the magnitude of $\vec{B}$ will be constant so that the integral in Eq. (5.46) reduces to $B \oint dl = B(2\pi r)$. Therefore,

$$B(2\pi r) = \mu_o I \frac{\pi r^2}{\pi R^2}.$$

Solving for B we get

$$B = \frac{\mu_o I r}{2\pi R^2} \quad \text{for } r \leq R.$$

For $r > R$ the enclosed current is just I so that

$$B = \frac{\mu_o I}{2\pi r} \quad \text{for } r > R.$$

The field magnitude increases in a linear way inside the conductor. Once outside, it will decay inversely with radial distance from the conductor. When $r = R$ the two expressions are equivalent. That is

$$B = \frac{\mu_o I}{2\pi R} \quad \text{for } r = R.$$

5.9 The Solenoid

In this section we will study a device that can be used to create a relatively strong and compact static magnetic field. This device has the desirable feature that the field can easily be turned on and off. We are referring to an item known as a *solenoid*. The solenoid uses a current to create a magnetic field but not a straight line current. We make a solenoid by taking a non-conducting hollow cylindrical tube and wrapping insulated wire around it. Such a device is depicted in Figure 5.16.

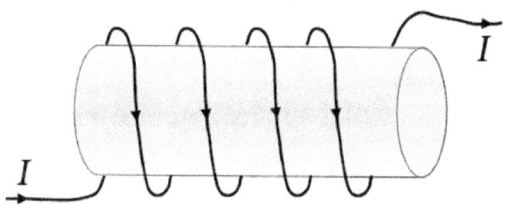

Figure 5.16: A solenoid. The arrow heads give the direction of positive current flow.

Once assembled, a current can be passed through the wire. Using the right hand rule for magnetic field lines it is seen that this current will create a concentrated magnetic field in the center of the cylinder directed from left to right. A convenient way to envision this is via a view of the solenoid split axially down the center. This is shown in Figure 5.17. where the current directions into or out of the plane of the page are denoted.

We can use this figure, along with Ampère's law, to determine a formula for the magnetic field magnitude in-

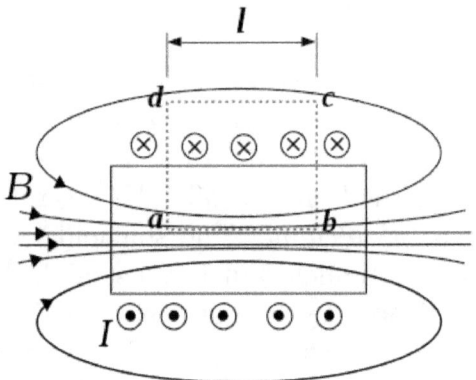

Figure 5.17: Split view of a solenoid with current flowing through its wires. Sections of some of the resulting magnetic field lines are sketched. The dashed rectangular box denotes a closed path used in the application of Ampère's law.

side the solenoid carrying current I. Consider the closed path shown in the figure: *abcda*. We assume that the magnetic field is nearly zero outside of the solenoid so that $B = 0$ along the path *cd*. Along the paths *bc* and *da* we assume that the magnetic field vectors are perpendicular to the path so that $\vec{B} \cdot d\hat{l} = 0$ along these two paths. That leaves only one section of the closed path that yields a non-zero value when applying Ampère's law, the section *ab*. Along this section $\vec{B}$ is in the same direction as $d\hat{l}$ so that Ampère's law becomes

$$B \int_{a-b} dl = \mu_o I_{enc} . \qquad (5.51)$$

The integral on the left of Eq. (5.51) will simply yield

5.9. THE SOLENOID

l. So that our result can be general, we assume that the closed path surrounds N loops (turns) of wire. Then, the current enclosed is NI. Using this in Eq. (5.51) and solving for B leads to

$$B = \frac{\mu_0 NI}{l} . \qquad (5.52)$$

A convenient way to write this formula is to let $n = N/l$ that is, n is the number of turns per unit length so that

$$B = \mu_0 nI . \qquad (5.53)$$

The simple solenoid is an example of an *electromagnet*. All such devices where an electric current is used to generate a magnetic field are referred to as electromagnets.

PROBLEMS

5.22 A magnetic dipole moment vector is given as $\vec{\mu} = 0.25\,\hat{x}$ Am^2. It is placed within a magnetic field given by $\vec{B} = 1.2\,\hat{x} - 3.5\,\hat{y}$ T. Compute the torque the field exerts on the dipole.

5.23 The magnetic moment of a compass needle is 0.1 Am^2. The magnetic field magnitude of Earth's magnetic field in North America is around 0.05 mT. If the needle, which can rotate in a plane parallel with Earth's surface, is initially at 30° from the geographic line of south to north what is the torque exerted by Earth's field on the needle?

5.24 Since a magnetic torque will cause an angular acceleration there must be some potential energy, U, associated with a magnetic dipole relative to a uniform magnetic field. This can be shown to be given by $U = -\vec{\mu} \cdot \vec{B}$. Compute the magnetic potential energy associated with the dipole and field of Problem 15.13.

5.25 Since a magnetic torque will cause an angular acceleration there must be some potential energy, U, associated with a magnetic dipole relative to a uniform magnetic field. This can be shown to be given by $U = -\vec{\mu} \cdot \vec{B}$. Compute the magnetic potential energy associated with the dipole and field of Problem 15.22.

5.26 Consider two parallel line currents, I_1 and I_2, a distance d apart. Show that the magnitude of the force per unit length, F_l, on current I_1 due to I_2 can be given by

$$F_l = \frac{\mu_0 I_1 I_2}{2\pi d}.$$

5.27 If a charged particle enters a uniform magnetic field with a velocity inclined at some angle θ from the field lines it will take a helical path along a field line. That is, it will not rotate in a plane since it will have a component of velocity that is in the same direction as the field line. Suppose that $\theta = 45°$, $|\vec{B}| = 0.33$ T and the charged particle is a proton and it enters the field with the speed 1.2×10^4 m/s. What will be the distance (pitch) between the loops in the helical path? That is, the distance between the loops along the direction of the field line.

5.28 Show that for an electron in circular orbit about a stationary proton the magnitude of the magnetic dipole moment can be written in terms of the angular momentum L as

$$\mu = \frac{qL}{2m},$$

where q is electron charge and m is electron mass.

5.9. THE SOLENOID

5.29 Estimate the magnetic moment for the electron orbiting the proton in the hydrogen atom. Let the radius of the orbit be 0.530 Å.

5.30 A toroidal solenoid is similar to a cylindrical solenoid only that it is in the shape of a toroid (a doughnut like shape). A rough sketch of such a solenoid is given below.

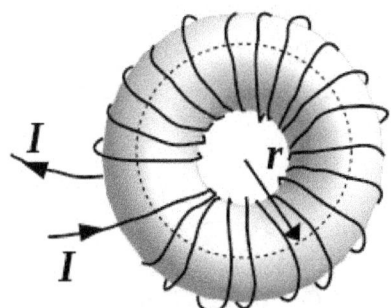

A DC current of I flows through insulated wire which is wrapped N times around the solenoid. Use Ampère's law to find an expression for the magnetic field magnitude inside the solenoid. By "inside" we mean within the hollow tube that is wrapped with insulated wire. (Hint: For the field inside, slice the toroid in half like a bagel then use a closed circular path of radius r).

5.31 A coaxial cable. A coaxial cable contains a central solid conductor in the shape of a cylinder surrounded by an insulating gap and then a metallic shielding in the shape of a cylindrical shell. Assume that the currents shown are uniformly distributed within the conductors. A section of such a cable is sketched in the figure below.

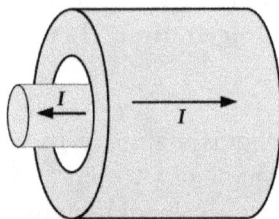

An end view of the cable is given in the figure below which gives the dimensions of the inner conductor and outer shield.

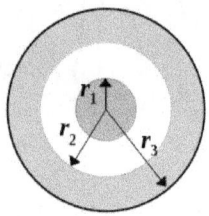

Use Ampère's law to find an expression for the magnetic field magnitude in terms of the radial distance R from the center of the cable in the four different regions: $R < r_1$; $r_1 \leq R < r_2$; $r_2 < R \leq r_3$ and $R > r_3$.

5.32 Recall that the curl of an electrostatic field is zero and since the curl of a gradient is also zero that meant we could write an electrostatic field in terms of the negative of the gradient of a scalar potential function. Well, the divergence of a static magnetic field is zero and the curl of a divergence is zero as well. This means we can write

5.9. THE SOLENOID

the magnetic field $\vec{B}$ in terms of some other vector field $\vec{A}$. That is

$$\vec{\nabla} \times \vec{A} = \vec{B} .$$

Notice how the divergence of both sides of this equation will be zero. We call $\vec{A}$ the *vector potential*. As with the scalar potential for electric fields, it gives us some analytic flexibility in that one can deal with $\vec{A}$ or $\vec{B}$ for a given situation.

The vector potential for a magnetic dipole, μ, is given by

$$\vec{A} = \frac{\mu_o \mu \sin\theta}{4\pi r^2} \hat{\phi} .$$

Find the magnetic field for this dipole then, verify that $\vec{\nabla} \cdot \vec{B} = 0$

Chapter 6

Magnetic Fields in Matter

6.1 Magnetization

Thus far we have considered the magnetic field in free space. Much like with the electric field, materials will have a specific reaction to being permeated by an external magnetic field so that these effects need to be accounted for within our models. Unlike in the case of the electric field, where the atoms or molecules of the dielectric are polarized in one direction, since atoms, molecules and electrons have magnetic moments, materials can respond in more than one way to the presence of a magnetic field.

We will see that the responses that materials have to external magnetic fields can be classified into two main categories: one in which the magnetic moments of the material will act to decrease the magnetic field magnitude within the material, the other where the reaction will act to increase the internal field magnitude. Materials that act to diminish the applied magnetic field are referred to as *diamagnetic*. Materials in which the applied field is increased inside, will for our purposes in this text, fall into to sub-catogories: *paramagnetic* and *ferromagnetic*.

The diamagnetic effect is the result of the external magnetic field interacting with the orbital magnetic moment of the atomic electrons. Imagine that the centripetal force generating the electron orbit was a Coulomb force between the electron and the nucleus. Now, when an external magnetic field of magnitude B is switched on there is an additional centripetal force of magnitude, qvB acting on the electron. This causes either the size of the electron orbit or its orbital speed to change or both. Either way, the observed effect is that there is a change in the atom's magnetic moment, $\Delta \mu$. This change creates a magnetic field at the atom (or molecule) that is oriented opposite that of the applied field. Therefore, **in diamagnetic materials the external magnetic field will be diminished within the material.**

Paramagnetism is a phenomenon caused by unpaired electrons. Electrons posses an unusual magnetic moment that only ever points in one of two directions. Theses states are referred to as spin up and spin down. You may recall from a chemistry or modern physics course that according to the Pauli principle, electrons within an atom tend to form pairs. When *paired*, one electron will be spin up, the other spin down. When in this paired state the magnetic moments of the electrons cancel. Since all atoms and molecules are to some degree diamagnetic, due to electron orbital motion, only atoms or molecules with an odd number of electrons will display paramag-

6.1. MAGNETIZATION

netism. Though all paramagnetic species will also have diamagnetic qualities, the paramagnetic interaction is the stronger of the two so that the net magnetic behavior of the material will be paramagnetic. Much like a compass needle, the external field will cause the unpaired electron spin to orient in a direction such that it reinforces the external field in the material. Therefore, **in paramagnetic materials the external magnetic field will be increased within the material.**

Ferromagnetism is a special case of paramagnetism. This effect is known to be displayed in the elemental metals Fe, Co and Ni. When the radius of the atom is large compared to the inter-atomic spacing, unpaired electrons on adjacent atoms can interact and align in the same direction. This seems to occur over regions in the metals that are called **domains**. These domains, can be viewed with an optical microscope. Though, Fe and Ni have an even number of electrons, many remained unpaired in the d orbitals. These issues at the atomic level need not concern us here. It is important for us to remember that as with paramagnetic materials, ferromagnetic materials act to reinforce any external magnetic field. However, unlike diamagnetism and paramagnetism, effects that vanish when the external field is removed, the ferromagnetic domains can remain aligned throughout the entire sample and produce a detectable external magnetic field.

As in our study of dielectrics it is useful here to define a collective quantity, like polarization, so as to account for the magnetic effects discussed above. Therefore, we define the *magnetization*, $\vec{M}$, as the magnetic moment per unit volume. However, unlike with polarization, the magnetization can be directed opposed to or in the direction of the applied field.

6.2 Bound Currents

As the effects of polarization could be dealt with via bound charges, the presence of a non-zero magnetization can be integrated into existing theory by relating it to bound currents. In this case there will be a bound volume current density, $\vec{J_b}$, and a bound surface current density, $\vec{K_b}$. Using the concept of a vector potential, introduced in Problem 5.33, one can show that

$$\vec{J_b} = \vec{\nabla} \times \vec{M} , \qquad (6.1)$$

and

$$\vec{K_b} = \vec{M} \times \hat{n} . \qquad (6.2)$$

Here, $\hat{n}$ is a unit vector at the surface and perpendicular to the surface of the magnetic material which always points away from the magnetized region. With these current densities known, either Biot-Savart or Ampère's law can be used to find the resulting magnetic field due to the magnetization.

Example 6.1

An upright solid cylinder with its axis aligned along the z direction has a frozen in magnetization given by $\vec{M} = K_o \hat{z}$ where K_o is a constant. Determine the bound surface and volume current densities, then use Ampère's law to estimate the magnetic field inside the cylinder.

Here $\vec{\nabla} \times \vec{M} = 0$ since $\vec{M}$ is uniform so that $\vec{J_b} = 0$.

The unit vector $\hat{n}$ will point in the $\hat{z}$ direction on either end where $\vec{M} \times \hat{n}$ will yield zero. On the sides $\hat{n} = \hat{r}$, so that,

$$\vec{K}_b = K_o \, \hat{z} \times \hat{r} = K_o \, \hat{\phi} \, .$$

The surface current density is on the surface, traveling in a counter clockwise direction when viewed in the $-z$ direction from the top of the cylinder.

We setup the closed loop for use with Ampère's law as shown in the figure below.

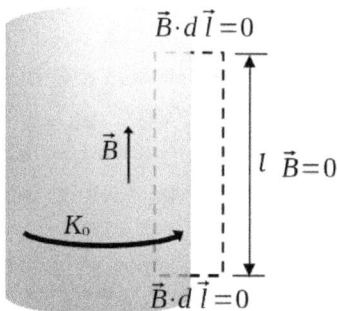

As with the solenoid, we assume that the magnetic field is zero outside of the cylinder and that $\vec{B} \cdot \vec{dl} = 0$ at both ends of our closed loop. Applying Eq. (5.46) yields one non-zero term on the inside of the cylinder. Since K_b is a current per distance, the current enclosed in the loop is $K_o l$ so that Eq. (5.46) gives,

$$Bl = \mu_o K_o l \text{ so that,}$$
$$B = \mu_o K_o \, .$$

This result is equivalent to that of the cylindrical solenoid where $B = \mu_o n I$. Here K_o takes the place of nI, as both have units of current per unit length. By the right hand rule, the interior field as a vector field is

$$\vec{B} = \mu_o K_o \, \hat{z} \ .$$

6.3 Ampère's Law in Materials

Having now introduced the concept of material magnetization it is important to re-evaluate the differential form of Ampère's law so as to account for the possibility of bound currents. Recall from Eq. (5.50), that the curl of the magnetic field was directly proportional to the enclosed current density. However, in that case the current density was due to an outside source and not a magnetization. As with free charge in a dielectric, we will refer to this current density as the free current density, $\vec{J}_f$. Now, Ampère's law in differential form is simply adjusted to account for the enclosure of both current types

$$\frac{1}{\mu_o} \vec{\nabla} \times \vec{B} = \vec{J}_f + \vec{J}_b \ . \qquad (6.3)$$

Using Eq. (6.1) in this and re-arranging leads to

$$\vec{\nabla} \times \left(\frac{\vec{B}}{\mu_o} - \vec{M} \right) = \vec{J}_f \ . \qquad (6.4)$$

We now define a new vector field $\vec{H}$ as

$$\vec{H} = \frac{\vec{B}}{\mu_o} - \vec{M} \ , \qquad (6.5)$$

6.3. AMPÈRE'S LAW IN MATERIALS

therefore
$$\vec{\nabla} \times \vec{H} = \vec{J}_f , \qquad (6.6)$$

So we see that the $\vec{H}$ field fulfills a role similar to that of the displacement field for electric fields in materials. That is, it is independent of the material medium which has become magnetized by an external field. Therefore, it is often easy to identify and write the expression for the $\vec{H}$ field for a magnetized material involving free currents as it depends only upon the free current, so that the bound currents need not be determined first.

In many older texts the $\vec{H}$ field is often referred to as the magnetic field while $\vec{B}$ is called the magnetic induction field. We will take the convention that $\vec{B}$ is the magnetic field and $\vec{H}$ is the magnetic displacement field or simply the H field.

Example 6.2

Consider the coaxial cable with two currents as shown in Problem 5.31. Find the H field in all regions.

Here the magnetization is zero so Eq. (6.5) gives $\vec{H} = \vec{B}/\mu_o$. Using the distances denoted in the figure from Problem 5.31, Ampère's law gives for $r < r_1$:

$$\vec{B} = \tfrac{\mu_o r I}{2\pi r_1^2} \hat{\phi} \text{ , so that}$$
$$\vec{H} = \tfrac{r I}{2\pi r_1^2} \hat{\phi} \text{ , for } r < r_1.$$

For $r_1 \leq r \leq r_2$:

$$\vec{B} = \tfrac{\mu_o I}{2\pi r} \hat{\phi} \text{ , so that}$$
$$\vec{H} = \tfrac{I}{2\pi r} \hat{\phi} \text{ , for } r_1 \leq r \leq r_2.$$

Now, for $r_2 \leq r \leq r_3$:

$$\vec{B} = -\frac{\mu_0 I}{2\pi r}\left(\frac{r^2-r_2^2}{r_3^2-r_2^2}\right)\hat{\phi}, \text{ so that}$$

$$\vec{H} = -\frac{I}{2\pi r}\left(\frac{r^2-r_2^2}{r_3^2-r_2^2}\right)\hat{\phi}, \text{ for } r_2 \leq r \leq r_3.$$

For $r > r_3$, I enclosed is zero so that $\vec{B} = 0$ and thus $\vec{H} = 0$.

You may recall from Eq. (5.42) that the divergence of a magnetic field is zero. So is the same true for the H field? We can examine this question bu considering the divergence of Eq. (6.5):

$$\vec{\nabla} \cdot \vec{H} = \vec{\nabla} \cdot \left(\frac{\vec{B}}{\mu_0} - \vec{M}\right). \tag{6.7}$$

Distributing through the del operator gives

$$\vec{\nabla} \cdot \vec{H} = \frac{\vec{\nabla} \cdot \vec{B}}{\mu_0} - \vec{\nabla} \cdot \vec{M}. \tag{6.8}$$

Since $\vec{\nabla} \cdot \vec{B} = 0$ we are left with

$$\vec{\nabla} \cdot \vec{H} = -\vec{\nabla} \cdot \vec{M}, \tag{6.9}$$

so that the divergence of $\vec{H}$ is zero only when the divergence of the magnetization is zero.

PROBLEMS

6.1 Use the fundamental theorem for curls on Eq. (6.6) to show that

$$\oint \vec{H} \cdot d\vec{l} = I_f ,$$

where, I_f is the free current.

6.2 Imagine that s solid sphere has a magnetization $\vec{M} = k_o r^2 \, \hat{r}$ where k_o is a constant. Verify that the corresponding current densities would be zero. What is $\vec{\nabla} \cdot \vec{H}$ in this case?

6.3 Suppose a solid cube of side length a has a permanent magnetization given by $\vec{M} = 20x \, \hat{z}$ A/m. Determine the bound current density.

6.4 Find the H field inside the toroidal solenoid discussed in Problem 5.30.

6.5 Use Eq. (6.6) to find the free current density in every region of the coaxial cable discussed in Example 6.2.

6.4 Linear Media

As with dielectrics, a good first approximation for the magnetization within a material is that it is directly proportional to the applied field. We refer to materials in which this is approximately the case as *linear magnetic materials*. However, as with most models of this type, one finds that the relation is only valid within a certain range of applied field magnitude.

It is tradition to write this relation as

$$\vec{M} = \chi_m \vec{H} , \qquad (6.10)$$

where χ_m is called the *magnetic susceptibility*. Using Eq. (6.5) in Eq. (6.9) gives

$$\vec{B} = \mu_o(\vec{H} + \vec{M}) . \qquad (6.11)$$

Using Eq. (6.9) in this leads to

$$\vec{B} = \mu_o(1 + \chi_m)\vec{H} . \qquad (6.12)$$

Now, we define the permeability μ as

$$\mu = \mu_o(1 + \chi_m) , \qquad (6.13)$$

so that

$$\vec{B} = \mu \vec{H} . \qquad (6.14)$$

Some authors define a *relative permeability* μ_r where, $\mu_r = (1 + \chi_m)$, so that Eq. (6.10) can be written as

$$\vec{B} = \mu_o \mu_r \vec{H} . \qquad (6.15)$$

As magnetization is the density of atomic magnetic moments with a material, its value can vary with the mass density of the medium. It is therefore common to define a *molar magnetic susceptibility*, χ_v, where

$$\chi_v = \chi_m V_m . \qquad (6.16)$$

Here V_m is the molar volume of the material, that is, the molecular/atomic weight divided by mass density.

6.4. LINEAR MEDIA

Table 6.1: Molar magnetic susceptibilities for various pure materials. Taken from, D.R. Lide, CRC Handbook of Chemistry and Physics, 88th ed., CRC Press, 2008.

Substance	χ_v 10^{-6} (cm^3/mol)
Al	+16.5
O$_2$ (s 54 K)	+10200
O$_2$ (l 90 K)	+7699
O$_2$ (g)	+3449
K	+20.8
B	-6.7
C (graphite)	-6.0
Cu	-5.46
Au	-28
KCl	-38.8
NaCl	-30.2

Due to the fact that paramagnetism and diamagnetism act in opposite directions on the applied field, there can be positive and negative magnetic susceptibilities. Values for the molar susceptibility of various pure substances are listed in Table 6.1.

An interesting result can be found by using Eq. (6.10) in Eq. (6.6). This gives

$$\vec{\nabla} \times \frac{\vec{M}}{\chi_m} = \vec{J}_f . \qquad (6.17)$$

Now, using Eq. (6.1) to replace $\vec{M}$ in this leads to

$$\vec{J}_b = \chi_m \vec{J}_f . \qquad (6.18)$$

208 CHAPTER 6. MAGNETIC FIELDS IN MATTER

Therefore, in a linear system, the bound current density is directly proportional to the free current density.

Example 6.3

Water has a molar susceptibility of -12.96 × 10^{-6} (cm^3/mol) at 293 K and 1.0 atm. However, water is observed to have a molar susceptibility of -13.09 × 10^{-6} (cm^3/mol) at 373 K. Assuming that the density of water varies in a linear way between these two temperatures use this data to estimate χ_m for water.

Write Eq. (6.16) in terms of the molecular weight of water W_m and the mass density of water at 293 K, ρ_1:

$$\chi_v = \chi_m \frac{W_m}{\rho_1}.$$

Solving this for χ_m and assuming that the density for water at 293 K is 1.0 g/cm^3, we get

$$\chi_m = \frac{(-12.96 \times 10^{-6} \text{ cm}^3/\text{mol})(1.0 \text{ g/cm}^3)}{18.0 \text{ g/mol}} = -0.72 \times 10^{-6}.$$

Since χ_m is assumed to be a constant we can use the above result to estimate the density for water near the boiling point ρ_2:

$$\rho_2 = \frac{18.0 \text{ g/mol} - 0.72 \times 10^{-6}}{-13.09 \times 10^{-6} \text{ cm}^3/\text{mol}} = 0.99 \text{ g/cm}^3.$$

6.4. LINEAR MEDIA

Example 6.4

An upright cylinder of water, with central axis long the z axis, is within a uniform field $\vec{H} = 0.12\ \hat{z}$ A/m. Find the magnetic field inside the cylinder, the magnetization and the bound currents.

Simply using Eq. (6.14), we get

$$\vec{B} = \mu(0.12)\ \hat{z}\ \text{A/m} = \mu_o(1+\chi_m)(0.12)\ \hat{z}\ \text{A/m}\ ,$$

Using χ_m for water found in Example 6.3, and the accepted value for μ_o, we get

$$\vec{B} = (4\pi \times 10^{-7})[1.0 - (-0.72 \times 10^{-6})](0.12)\ \hat{z}\ \text{T} = 1.51 \times 10^{-7}\ \hat{z}\ \text{T}\ .$$

We use Eq. (6.10) to get the magnetization:

$$\vec{M} = (-0.72 \times 10^{-6})(0.12)\ \hat{z}\ \text{A/m} = -8.64 \times 10^{-8}\ \hat{z}\ \text{A/m}.$$

We use Eq. (6.2) to get the bound surface current density $\vec{K}_b$ with $\hat{n} = \hat{r}$. Note that, $\hat{z} \times \hat{r} = \hat{\phi}$ therefore

$$\vec{K}_b = 8.64 \times 10^{-8}\ \hat{\phi}\ \text{A/m}.$$

Now to get the volume current density $\vec{J}_b$ we need the curl of $\vec{M}$ in cylindrical coordinates. From an Appendix we find that $\vec{\nabla} \times \vec{M} = 0$ in this case so that

$$\vec{J}_b = 0\ .$$

6.5 Magnets

Previously in this chapter we discussed the phenomenon of ferromagnetism. In this section we will take a closer look at how ferromagnetic materials behave in the presence of external magnetic fields. Also, the assumption that magnetic materials are linear in the response to a magnetic field is also examined further in this section. We will find that linearity is often a good approximation only when the magnitude of the external field is relatively small. This fact will be revealed when we consider a ferromagnetic material, such as Fe, within a magnetic whose magnitude and direction are varied.

Fe is one of the few pure materials where electron spins, between neighboring atoms within the material, can become collectively aligned in such a way as to generate a significant magnetic field within particular regions or *domains*. In its natural state, there is no net external magnetic field around an iron sample as the net dipole moment for each domain will be oriented in directions which tend to cancel one another. However, if an iron sample is placed in an external uniform magnetic field of sufficient magnitude then, the moments of each domain will align with the field lines of the external field–similar to the way in which the lone magnetic dipole moment aligned with a field line in an external magnetic field as discussed in a previous chapter. Once the external field is removed, the dipole moments of the domains will remain aligned and the sample will have a detectable magnetic field surrounding it. This process is referred to as *magnetization*. We say the material has become *magnetized* and henceforth refer to the sample as a *magnet*. If it is in the shape of a rectangular slab, as depicted in Fig. 6.1,

6.5. MAGNETS

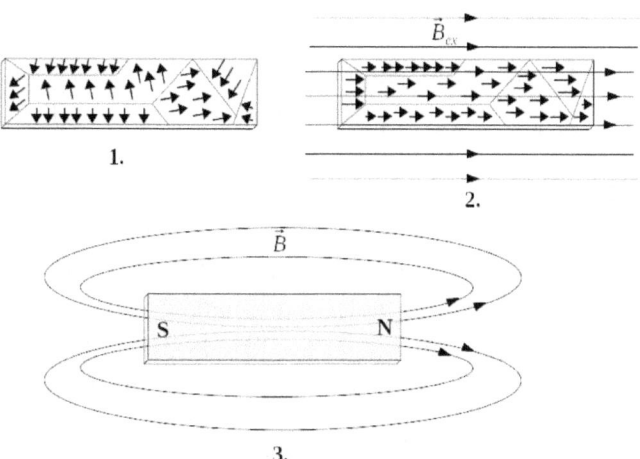

Figure 6.1: **Magnetization. 1.** In its natural state, iron has magnetic dipoles aligned within domains. An abstracted view of some of these dipoles and domains are depicted in this figure. The net magnetic moment of each of these domains tend to cancel one another so that the sample has no detectable magnetic field. **2.** When placed within an external magnetic field, the dipoles within the sample will align with the lines of the external field. **3.** Once removed from the external field, the dipoles remain aligned and the sample, now called a bar magnet, has a detectable magnetic field.

we call it a bar magnet. This magnetization process is illustrated in three steps in Fig. 6.1.

Ferromagnetic materials display a phenomenon known as hysteresis. In the process discussed in Figure 6.1 once the sample became magnetized, the external field was removed. However, interesting properties are revealed when we consider the magnetization process in greater detail. For example, let us consider what might occur when the magnitude of the external field is increased from zero to some higher value. Then, what happens when the field is reduce to zero and then switches directions? Such observations indicate that ferromagnetic materials display a hysteresis effect in that the value of the magnetization depends upon the magnetization in the immediate past.

To see how all of this works out, consider the idealized sketch of the magnitude of magnetization versus magnitude of $\vec{H}$ in a ferromagnetic sample as depicted in Figure 6.2. Initially, at point a, the sample is unmagnetized and the external field is zero. Then, the external field strength is increased and the magnetization increases according to the curve going from point a to point b.

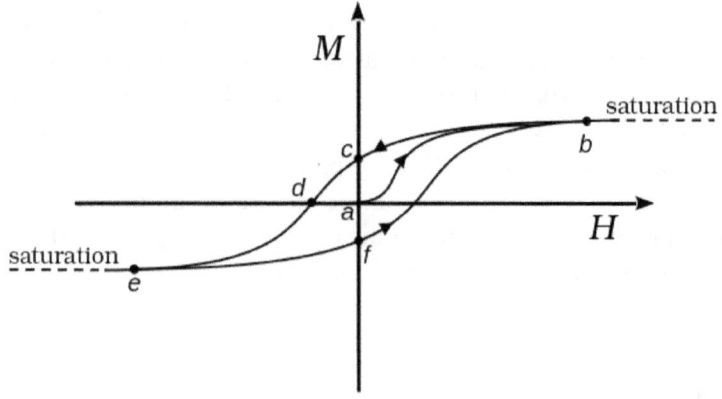

Figure 6.2: Idealized magnetization vs. external field for a ferromagnetic sample displaying the hysteresis effect.

Note that along this path, M versus H is only approximately linear for small values of H. As the field strength is further increased we see that the magnetization reaches a terminal value beyond which further increases in H do

6.5. MAGNETS

not yield additional increases in M. Simply put, all of the atomic moments that can align with the field have becomes aligned and no further increase in M can occur. We say that the material has reached the magnetic *saturation* point.

Now, suppose we decrease H. Rather than the magnetization following the curve back to point a it will have values along the line to point c. That is, the material will retain its magnetization and behave as a permanent magnet. Now suppose we reverse the direction of the external field and begin slowly increasing its magnitude. The magnetization versus H path will be along the line from point c to point d. That is, due to the sample's history, a non-zero field is required to eliminate the magnetization within the sample. Continuing to increase the field magnitude, as with the field directed in the opposite direction, we will reach a saturation point, point e, beyond which the magnetization in the sample cannot be increased. The cycle is concluded by reducing the field magnitude until point f is reached where the sample retains a residual magnetization, as at point c, only in the opposite direction. That is, the poles have been flipped.

The magnetic sample discussed above is referred to as a *hard ferromagnet*. Hard ferromagnets retain their magnetization when the external field is removed as seen in Figure 6.2 at points f and c, whereas *soft ferromagnets* lose their magnetization once the field is removed. Examples of hard ferromagnetic materials are the Fe, Al, Co, so-called, *Alinco* alloys and mild chromium steel. An example of a soft magnetic material would be Mn-Zn ferrite.

Even for a hard magnetic material the magnetization

condition is not permanent. For temperatures above absolute zero, vibrations at the atomic level tend to disorder the magnetic dipole moments over time. As might be expected then, increasing temperature tends to erase the magnetization condition faster. At temperatures at or above the *Curie temperature*, the magnetization of the sample vanishes and cannot be reconstituted. For Fe, the Curie temperature is around 1043 K. Also, mechanical shock can disrupt the sample's magnetization.

It should be noted that the hysteresis plot given in Figure 6.2 is an idealization. Real M versus H plots are often more irregular in appearance yet possess the same essential features discussed above.

PROBLEMS

6.6 Consider the coaxial cable from Problem 5.31. Assume that the region $r_1 < r < r_2$ is filled with a linear magnetic material of susceptibility χ_m. Find $\vec{H}$ in all regions and the bound current densities.

6.7 At the flat surface of a magnetic material there is an angle of 30° between the surface and $\vec{H}$. Find the magnitude of the bound surface current density.

6.8 In an upright cylinder of magnetic material of radius R, there is a free current density $\vec{J}_f = J_o\,\hat{z}$. Find possible expressions for the magnetization.

6.9 Get the mass density at room temperature (or at the temperature listed) for the substances listed in Table 6.1 then, use their molecular weights to compute χ_m in each case.

6.5. MAGNETS

6.10 A solid cylinder of graphite of radius 2.0 cm, has a straight line current of 2.0 A running along its central axis. Find expressions for $\vec{B}$ and $\vec{H}$ in all regions, Then, find the magnetization and the bound current densities.

6.11 A permanent magnet is in the shape of a solid toroid with a small air gap as shown in the figure below. The magnetization uniform and in a circumferential direction in the material. Find expressions for the Magnetization, $\vec{M}$ and the $\vec{H}$ and $\vec{B}$ fields inside the magnet and in the air gap.

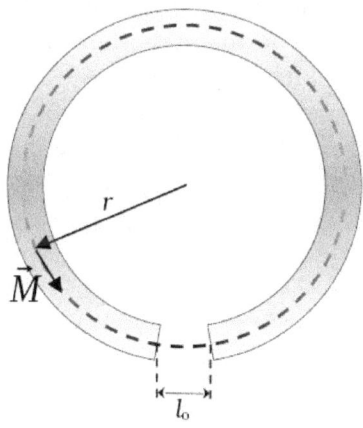

6.12 A straight, infinite DC current I, is directed along the central axis of a solid cylinder made of non-conducting linear magnetic material. Find $\vec{M}$, $\vec{H}$ and $\vec{B}$ in all regions. Also find the bound current densities.

6.13 In some region there is a magnetization given by $\vec{M} = 5y\,\hat{x} + 6x^2\,\hat{y}$ A/m. Find the corresponding bound current density $\vec{J_b}$. Compute $\vec{\nabla}\cdot\vec{H}$. Is this a case where $\vec{\nabla}\cdot\vec{H} = 0$?

16.14 An upright cylinder of graphite, with central axis long the z axis, is within a uniform field $\vec{H} = 0.25\,\hat{z}$ A/m. Find the magnetic field inside the cylinder, the magnetization and the bound currents.

16.15 In a strange 2D magnetic material $\vec{\nabla} \cdot \vec{H} = M_o y^2$, where M_o is some constant. If the magnetization is given by $\vec{M} = m(y)(\hat{x} + \hat{y})$ determine the function $m(y)$.

16.16 Consider a uniformly magnetized ($\vec{M} = M_o\,\hat{z}$) solid sphere of radius r. The magnetic field magnitude along the z axis outside the sphere will be $B_1 = (2\mu_o M r^3)/(3z^3)$. The magnetic field magnitude along the z axis inside the sphere will be $B_2 = (2/3)\mu_o M$. Find $\vec{H}$ inside and outside the sphere.

Chapter 7

Faraday's Law

7.1 Magnetic Flux

The idea of the flux of a vector field has been covered in a previous chapter and in particular, the flux of the electric field was of importance when discussing Gauss's law. We now wish to revive this notion for the case of magnetic fields. We therefore define the flux, Φ_B, of a magnetic field, $\vec{B}$, through some surface, S, as

$$\Phi_B = \int_S \vec{B} \cdot d\vec{A} . \qquad (7.1)$$

The SI unit for magnetic flux will T m^2 which goes by the special name, Weber (Wb).

Example 7.1

a) A uniform magnetic field is given by $\vec{B} = 0.2\,\hat{z}$ T. What is the flux of this field through a square area lying on the xy

plane of width 0.25 m?

b) A uniform magnetic field is given by $\vec{B} = 0.1\,\hat{x} + 0.2\,\hat{z}$ T. Determine the flux of this field through the surface in part a).

c) A nonuniform magnetic field is given by $\vec{B}(x) = 0.1\,\hat{x} + 0.2x\,\hat{z}$ T. Determine the flux of this field through the surface in part a).

a) Here, the field is uniform and perpendicular to the area in question so that Eq. (7.1) simplifies to:

$$\Phi = (0.2)(0.25)(0.25) = 0.0125 \text{ Wb}.$$

b) In this case, the field vectors are uniform across the surface but not normal to the surface. We could use a simplified version of Eq. (7.1), $\Phi = BA\cos\theta$, if the angle θ were known. However, it is instructive to use Eq. (7.1) instead. We have that $d\hat{A} = dxdy\,\hat{z}$. We first compute the required dot product:

$$(0.1\,\hat{x} + 0.2\,\hat{z}) \cdot dxdy\,\hat{z} = (0.2)dxdy.$$

We put it all together to get

$$\Phi = (0.2) \int_0^{0.25} \int_0^{0.25} dxdy = (0.2)(0.25)(0.25) = 0.0125 \text{ Wb}.$$

c) Lastly, we have a situation where the field varies with position. We use Eq. (7.1) again. Computing the dot product leads to

$$(0.1\,\hat{x} + 0.2x\,\hat{z}) \cdot dxdy\,\hat{z} = 0.2x\,dxdy.$$

Computing the flux gives

$$\Phi = (0.2) \int_0^{0.25} \int_0^{0.25} x\,dxdy = (0.2)(0.25)\frac{(0.25)^2}{2} = 1.56 \times 10^{-3} \text{ Wb}.$$

7.2 Electromagnetic Induction

Having been introduced to the concept of magnetic flux, we are now in a position to discuss one of the powerful laws of electricity and magnetism that resulted from work on the subject in the 19th century–*Faraday's law*.

In the previous chapter we learned that a current generates a magnetic field. Perhaps you wondered if the inverse were true? That is, can a field generate a current in a conductor? We will learn in this section that in certain cases the answer is yes.

Consider a circular conducting loop that inscribes an area A. It is an interesting fact of nature that if there is a magnetic flux through this area that changes over time t, it will induce a current in the loop. If there is an induced current, there must be an induced electric potential difference or voltage. When a voltage is induced via a changing magnetic flux it is referred to as an *electromagnetic force* or *emf*. This is simply a tradition. We could just as well call the induced potential a voltage but will continue to refer to an induced voltage as an emf in this chapter. Let the symbol $\mathscr{E}$ denote an emf. A single conducting loop with changing magnetic flux through the area it inscribes is depicted schematically in Figure 7.1.

Faraday's law gives a quantifying relationship for this effect. It states that for the a single current loop

$$\mathscr{E} = -\frac{d\Phi}{dt} . \qquad (7.2)$$

The presence of the minus sign in the above expression is a convention meant to convey the fact that the induced voltage, and thus induced current, acts to oppose the change in flux; a process that will be discussed further

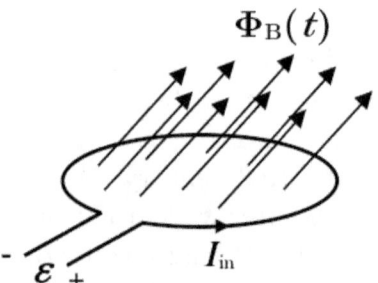

Figure 7.1: A changing magnetic flux Φ through the area inscribed by a current loop induces current I_{in} and emf $\mathcal{E}$.

when a corollary to Faraday's law, Lenz's law, is presented. Regardless of the sign, one can depend upon Eq. (7.2) to yield the correct absolute value for the induced voltage.

Suppose the conductor were coiled into N loops? Then, Faraday's law is easily extended as

$$\mathcal{E} = -N\frac{d\Phi}{dt}. \tag{7.3}$$

Using the formal definition of flux from Eq. (7.1), we can write Faraday's law in terms of the magnetic field and the inscribed area as

$$\mathcal{E} = -N\frac{d}{dt}\int_S \vec{B} \cdot d\hat{A}. \tag{7.4}$$

It is important to remember that the magnetic flux depends upon two things, the field and an area. This is why in Eq. (7.4) we see that an emf can be induced either by

7.2. ELECTROMAGNETIC INDUCTION

a magnetic field that changes over time, an inscribed area that changes over time or both. Often, we will consider cases where the initial and final values of a magnetic field magnitude and/or an area, are known over a particular time interval Δt. In this case, Faraday's law can be approximated as

$$\mathcal{E} = -N\frac{\Delta(BA)}{\Delta t}. \qquad (7.5)$$

The process of generating a voltage in a circuit via a field, either electric or magnetic, in which there is no direct contact between the field source and the circuit is referred to as *electromagnetic induction*.

Example 7.2

A uniform magnetic field is directed perpendicular to the circular conducting loop shown in the figure below. Here $R = 20.0\ \Omega$ and the loop has a radius of 10 cm.

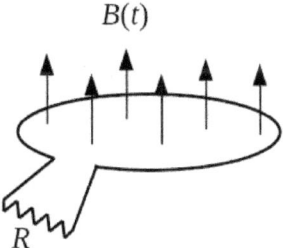

Figure 7.2: A changing magnetic field passing through the circular area inscribed by a conducting loop.

a) If the magnetic field magnitude is initially 1.0 T and then over a time period of 0.25 s increases to 2.0 T, what is the current through the resistor?

b) Suppose the loop shown above lies in the xy plane and that the magnetic field is the same throughout the loop but varies with time as $\vec{B} = B_o t \,\hat{z}$ T, where B_o is a constant. Derive an expression for the current through the resistor.

a) Here, we simply use Eq. (7.5) to get the emf and then use Ohm's law to get the current.

$$\mathscr{E} = -\frac{(2.0-1.0)(\pi(0.1)^2)}{0.25} = 0.126 \text{ V}.$$

$$I = \frac{\mathscr{E}}{R} = \frac{0.126}{20} = 6.3 \text{ mA}.$$

b) In this case, $\vec{B}$ does not vary with position and is everywhere perpendicular to the inscribed area so that $\vec{B}$ is factored out of the integral of Eq. (7.4). The remaining integral then simply yields the area of a circle so that Eq. (7.4) reduces to

$$\mathscr{E} = -\frac{d}{dt}(B_o)t\pi r^2 = -(B_o)\pi r^2.$$

Now by Ohm's law

$$I = \frac{-B_o \pi r^2}{R}.$$

PROBLEMS

7.1 A magnetic field in cylindrical coordinates is $\vec{B} = B_o \phi^2 \,\hat{z}$ T, where B_o is a constant. Find the flux of this field through a circle of radius r on the xy plane.

7.2 A magnetic flux within a closed conducting loop varies with time t as $\Phi(t) = \Phi_o e^{-kt^2} \sin(\omega t)$, where Φ_o, k and ω are constants. Find the induced emf. Sketch a plot of this induced voltage over time.

7.3 A loop of N turns has a time changing magnetic flux through its inscribed area given by $\Phi(t) = \Phi_o \sin(\omega t)$ where Φ_o is a constant and ω an angular frequency. Derive an expression for the emf induced in the loop.

7.4 A conducting loop of N turns lies within a uniform magnetic field as given by $\vec{B} = 0.5\,\hat{y} + 0.4\,\hat{z}$ T. The loop, which lies in the xy plane, is changing in shape over time so that the area it inscribes is given by $A(t) = A_o t^2$ m^2 where A_o is a constant. Derive an expression for the emf induced in the loop.

7.3 Induction in a Moving Conductor

Interestingly, an emf can be induced within a conductor simply by moving it through a static and uniform magnetic field. Consider a conducting rod of length l. Imagine that the rod moves with velocity $\vec{v}$ through a uniform magnetic field as depicted in Figure 7.3.

Since certain electrons are free to move in a conductor, by the right hand rule for the cross product, it can be shown the the magnetic force will move electrons to the bottom of the rod leaving excess positive charge at the top. This sets up an electric field, and thus a voltage or

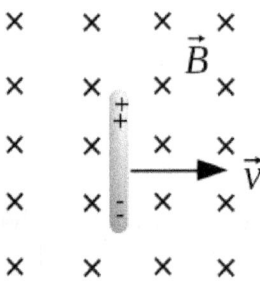

Figure 7.3: A conducting rod moving through a uniform magnetic field.

emf in the rod. Since the rod has length l we can use a result from Example 3.3 to write

$$E = \frac{\mathcal{E}}{l} . \qquad (7.6)$$

This situation will be stable as long as the velocity and the field remain constant. Therefore it must be that the magnetic force is balanced by the Coulomb force so that $E = vB$, where v is the magnitude of the rod's velocity. Using this fact in Eq. (7.6) and solving for the emf we get

$$\mathcal{E} = Blv . \qquad (7.7)$$

Now, consider the same rod in an identical magnetic field only now the rod slides along a u-shaped conductor as shown in Figure 7.4. During the time period dt there has been a change in area inscribed by the loop of $lvdt$. Since the magnetic field magnitude remains con-

7.3. INDUCTION IN A MOVING CONDUCTOR

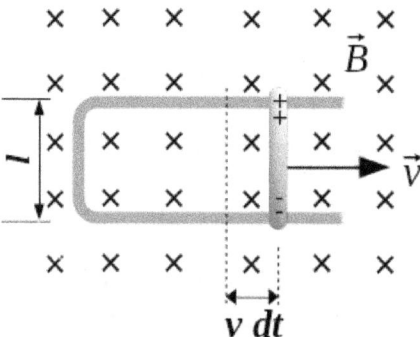

Figure 7.4: A conducting rod along a u-shaped conductor at speed v in a uniform magnetic field.

stant, from Eq. (7.3) we get

$$\mathcal{E} = -\frac{Blvdt}{dt} = -Blv \ . \tag{7.8}$$

This is identical to the result obtained for the rod by itself only in this case a current will flow in the loop.

Energy is conserved, so the induced emf must come from the mechanical work required to move the rod. It is useful to then relate the quantity of power, P, with the parameters involved in this problem. In this case, the induced current will flow counter-clockwise through the above loop. Therefore, by the right hand rule, there must be a magnetic force that opposes the direction of motion. That is, we must apply at least a force equal in magnitude to this magnetic force to maintain the motion. From Eq. (5.34) this force must be lIB. Since the speed is constant, it must be that the distance traversed by the rod in time t must be vt so that the work, W, done is

$$W = lIBvt . \tag{7.9}$$

The instantaneous power, P, is given by $P = dW/dt$ so that we have

$$P = lIBv . \tag{7.10}$$

Rather than use current, let us write Eq. (7.10) it in terms of the resistance, R, of the loop. Using Ohm's law in Eq. (7.10) leads to

$$P = \frac{lBv\mathscr{E}}{R} . \tag{7.11}$$

To eliminate the emf, a quantity difficult to measure in this case, use Eq. (7.8) in Eq. (7.11). Since a positive value for power is required we ignore the minus sign in Eq. (7.8) with the final result being

$$P = \frac{(Blv)^2}{R} . \tag{7.12}$$

This power originates from the mechanical force pulling the sliding rod and is then dissipated as heat when current flows through the circuit.

Example 7.3

A conducting rod of length l moves along a u-shaped conductor within a uniform magnetic field as depicted in Figure 7.4. Starting from rest the rod moves at a constant speed of 1.2 m/s to the right. Initially, the area inscribed by the conducting loop is l^2. Here, $l = 12$ cm, $B = 5.5$ T and $R = 2.0$ Ω. What is the induced emf and what is the total power delivered

7.4. SELF INDUCTANCE

to the circuit by this motion over a time period of 10.0 s?

To determine the induced emf, we need to compute the change in magnetic flux. This requires knowledge of the distance traversed by the rod during the 10 s time interval. From kinematics this is simply

$$x = vt = (1.2)(10.0) = 12.0 \text{ m}.$$

So, the change in area was

$$\Delta A = (l^2 + 12l) - l^2 = 12l = 12(0.12) = 1.44 \text{ m}^2.$$

Now, the induced emf is computed using Eq. (7.5):

$$\mathcal{E} = -\frac{B\Delta A}{\Delta t} = -\frac{(5.5)(1.44)}{10.0} = -0.79 \text{ V}.$$

The power, or rate of energy input, is independent of the time period so we simply use Eq. (7.12):

$$P = \frac{[(5.5)(0.12)(1.2)]^2}{2.0} = 0.32 \text{ W}.$$

7.4 Self Inductance

We have discussed the fact that an emf can be induced in a conducting loop given a change in the magnetic flux through the loop; this changing flux being due to a changing magnetic field or a change in the area inscribed by the loop. However, since magnetic fields are generated by currents it ought to be that a time varying current can be used to induce an emf in a conducting loop. A neat example of this effect is when a conducting loop, which carries a time dependent current,

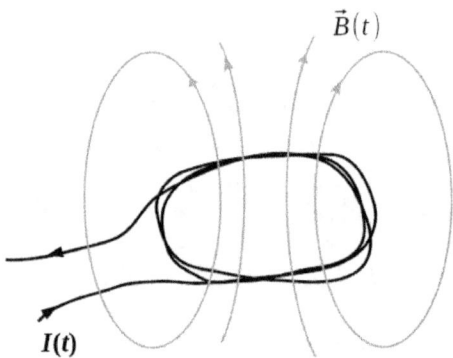

Figure 7.5: A coil with N turns through which a time varying current flows.

induces an emf upon itself. To understand how this occurs consider Figure 7.5.

Here, a time varying current passes through a conducting loop of N turns. By the right hand rule, this generates a magnetic field around the loop. In our study of Ampère's law we learned that magnetic fields are typically directly proportional to some current. This implies that the flux through the loop in Figure 7.5 must be changing over time. Therefore, an emf is induced. This type of induction is referred to as *self inductance*.

We can use Eq. (7.5) to describe the emf induced in this case and in doing so be introduced to an important new physical quantity. Consider some change in the current through the coil in Figure 7.5 given by ΔI over the time period Δt. B is directly proportional to the current so we write that $B = \mathscr{C}\Delta I$ where $\mathscr{C}$ is some constant.

7.4. SELF INDUCTANCE

Putting all of this into Eq. (7.5) we get.

$$\mathscr{E} = -\frac{\mathscr{C}\Delta I A}{\Delta t} \ . \tag{7.13}$$

Now, it is convenient to define $L = \mathscr{C}A$ where L is called the self-inductance or simply the *inductance* of the coil. L has the SI unit of the Henry (H). Letting $\Delta t \to 0$ we get the instantaneous self induced emf in the coil as

$$\mathscr{E}(t) = -L\frac{dI}{dt} \ . \tag{7.14}$$

A coil of this type is referred to as an *inductor* and has useful properties when used within an electric circuit. The schematic symbol for an inductor is shown below in Figure 7.6.

Figure 7.6: Circuit symbol for a coil or inductor.

Example 7.4

Assume that the magnetic field in the interior of an inductor can be given by the expression for the field inside a cylindrical air-core solenoid. Derive an expression for L in terms of the parameters of the solenoid.

We get from Eq. (5.53) that for a solenoid

$$B = \mu_o n I \ ,$$

where n is the number of turns per unit length. We use this in Eq. (7.5) to get

$$\mathcal{E} = -N\frac{\mu_0 n \Delta I A}{\Delta t} \ .$$

Here A is the constant cross-sectional area of the coil. Letting $\Delta t \to 0$ and factoring leads to

$$\mathcal{E} = -N\mu_0 n A \frac{dI}{dt} \ .$$

On comparing this with Eq. (7.14) we see that

$$L = N\mu_0 n A \ .$$

Since $n = N/l$, where l is the length of the solenoid we finally get

$$L = \frac{\mu_0 N^2 A}{l} \ . \tag{7.15}$$

7.5 Lenz's Law

In the preceding section it was perhaps not clear what effect the induced current had upon the current that created it. That is, did the newly induced current move in the same direction as the applied current $I(t)$ or does it oppose it? It turns out that the induced current will act to oppose the change in the applied current. One could determine all of this by a careful application of Faraday's law but it is easier to apply another famous law of physics to give the direction of the induced current. This principle is referred to as *Lenz's law*.

This rule is easy to state but requires some practice in order to be applied correctly. Lenz's law states that

7.5. LENZ'S LAW

The induced current will be in a direction so as to oppose the change in flux that induced it.

Let us consider an example where the direction of induced current is determined by application of this rule. Consider Figure 7.7 where a bar magnet is shown configured along the vertical and held above a conducting ring that lies in the xy plane.

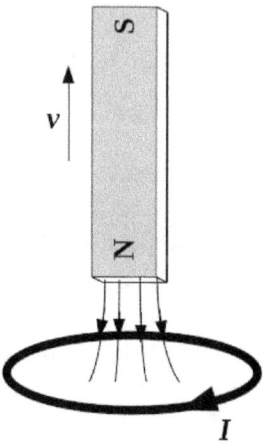

Figure 7.7: A bar magnet is held above a conducting ring and then is pulled upwards. A current is induced in the conducting ring below.

Imagine that the bar magnet is then pulled upward. There will be a change in magnetic flux through the ring and thus an induced current. Lenz's law can be used to give the direction of the induced current. Sections of a few of the magnetic field lines coming from the bar magnet are sketched in the figure. Note that they are initially directed downward through the loop. When the magnet is pulled upwards the intensity of the downward directed magnetic field lines through the loop will decrease. By Lenz's law the induced current will be in a direction that will oppose this change. That is, the induced current will act to form more downward directed magnetic field lines to replace the loss. We see that by the right hand rule for magnetic fields that current will need to flow in the clockwise direction, as viewed looking down from the magnet.

Example 7.5

In the solenoid shown below the current is in the direction indicated and increasing. A circular conducting loop is to the right of the solenoid and the plane of its inscribed area is parallel to the plane of the circular cross-section of the solenoid. The viewing perspective desired for the study of this problem is denoted by an eye. In what direction is the induced current with respect to this point of view?

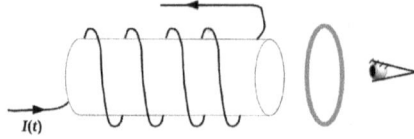

Figure 7.8: A time dependent current is increasing through a solenoid. A current is induced in the conducting ring at the right.

7.5. LENZ'S LAW

By following the path of the current, we find the direction that it takes around the solenoid through the winding. Then, we use the right hand rule to find the direction of the magnetic field lines inside the solenoid. These results are sketched in the figure below.

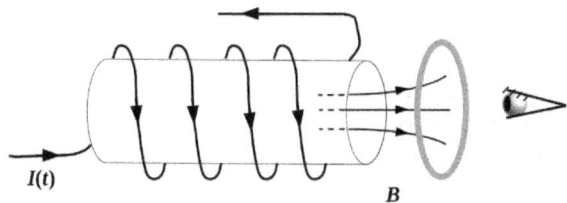

Figure 7.9: Direction of current and field lines denoted for the situation from Fig. 16.10.

This leads to magnetic field lines pointing from left to right through the conducting ring. Since the current is increasing the magnetic field and thus the flux will be increasing in the left to right direction through the ring. By Lenz's law the induced current will act to oppose this change. Therefore, the induced current will produce field lines that point from right to left through the ring. Using the right hand rule, from the point of view of the eye, we find that the induced current will rotate clockwise.

PROBLEMS

7.5 A conducting rod of length l moves along a u-shaped conductor in a uniform magnetic field as shown in Figure 7.4. With a field magnitude B and speed v and a loop resistance of R show that the induced current I is $I = Blv/R$. Now, while keeping l and v the same, if the field magnitude is doubled while the resistance is halved, what is the new induced current?

7.6 An inductor is required that has an inductance of 1.0 mH. You have a cylinder of circular cross section of radius 0.2 cm and length 1.2 cm. How much insulated wire is required?

7.7 Derive a formula for the self inductance, L, for the air-core toroidal solenoid considered in Problem 5.30. Let the cross-sectional area of the solenoid core be A. (Careful, here the field varies with position.)

7.8 In the figure below a conducting loop is being pulled from an area of uniform magnetic field into a region with no field. In which direction is the induced current in the loop?

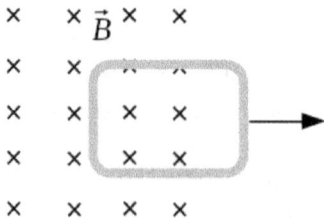

Figure 7.10: A conducting loop being pulled from within a uniform magnetic field.

7.6 Energy in a Magnetic Field

Recall, that a charged capacitor can be thought of as storing a potential energy, U. The same is true for an inductor with a current running through it. We say that the capacitor stores energy in an electric field whereas the inductor stores energy in a magnetic field. Previously, we derived a formula for the energy stored in a capacitor in terms of its capacitance. Let us now find an analogous formula for the inductor in terms of its inductance, L.

Recall from Eqs. (5.14) and (5.17) that power P in an electric circuit can be given as the product of a current and a voltage. Therefore, when an inductor has a time varying current flowing through it the instantaneous power can be given by

$$P = LI\frac{dI}{dt}, \qquad (7.16)$$

where we have used Eq. (7.14). Then, the differential for work W must be

$$dW = Pdt = LIdI. \qquad (7.17)$$

The total energy, U, stored in the inductor must equal the total work done on bringing the current through the inductor from zero to the final steady state value of I so that

$$U = \int_0^I LIdI. \qquad (7.18)$$

Carrying out the integration above leads to

$$U = \frac{1}{2}LI^2. \qquad (7.19)$$

Using Eq. (7.15) in this gives

$$U = \frac{\mu_o N^2 A I^2}{2l}. \qquad (7.20)$$

Using the expression for the field in the interior of a soleniod in Eq. (7.20) yields

$$U = \frac{1}{2}\frac{lAB^2}{\mu_o}. \qquad (7.21)$$

If we consider some unit volume of space within the region of a uniform field of magnitude B Eq. (7.21) leads to the energy density u_o in the uniform magnetic field:

$$u_o = \frac{B^2}{2\mu_o}. \qquad (7.22)$$

7.7 Mutual Inductance

In this section we will study one of the most important applications of Faraday's law. Previously, we considered how a changing current through a wire loop will induce a voltage on the loop itself–so called self inductance. Suppose now there are two conducting insulated wire loops that make no direct contact with one another but are near one another. By near we mean that they are close enough so that the magnetic field from one, can be easily sensed by the other. For example, the two coils might be on top of one another. Two such coils are depicted in Figure 7.11.

We can use Faraday's law to derive an expression for the voltage induced in each coil and in doing so arrive at an important relation for a device that exploits the

7.7. MUTUAL INDUCTANCE

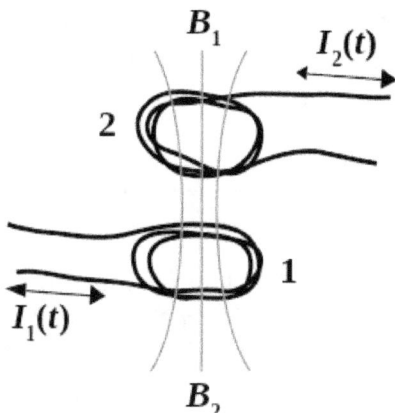

Figure 7.11: A time varying current in coil 1 induces a time varying current in coil 2.

inductive effect between two coils. Let us assume that coil 1 is of N_1 turns while coil 2 is composed of N_2. The changing current at coil 1 induces a voltage in coil 2 we denote as V_2. From Faraday's law we have that

$$V_2 = -N_2 \frac{d\Phi_2}{dt} . \qquad (7.23)$$

Let the area, A, inscribed by both coils be constant and equal. Let the length of both coils be given by l. Now assume the magnetic field magnitude generated by a current through the coil is equivalent to that of a solenoid. Using Eq. (5.53) in Eq. (7.23) we get

$$V_2 = -N_2 \frac{d(\mu_o n_1 I_1 A)}{dt} . \qquad (7.24)$$

Note here that the flux through coil 2 is due to the field of coil 1 so that $n_1 = N_1/l$ gives the turns per unit length of coil 1. In Eq. (7.24), only the current varies with time so that we can simplify things as

$$V_2 = -\frac{N_2 N_1 \mu_o A}{l}\left(\frac{dI_1}{dt}\right). \qquad (7.25)$$

This induced voltage in the second coil creates a time varying current through it we label as I_2. The field due to I_2 induces the voltage V_1 in coil 1. Using the same argument as above, we get an analogous formula for the voltage induced in coil 1, V_1 due to coil 2:

$$V_1 = -\frac{N_2 N_1 \mu_o A}{l}\left(\frac{dI_2}{dt}\right). \qquad (7.26)$$

It is seen that both Eqs. (7.25) and (7.26) have the same constant of proportionality which we label as M, the *mutual inductance*. So, for this two coil system, the mutual inductance is given by

$$M = \frac{N_2 N_1 \mu_o A}{l}. \qquad (7.27)$$

As with self inductance, the SI unit is the Henry.

Now, we make an important assumption about the two coils that allows us to derive an additional and quite useful formula. It is assumed that the fields perfectly couple between the coils so that is at all times $B_1 = B_2$. Letting the field magnitude be given by B allows Eqs. (7.24) and (7.25) to be written as

$$V_2 = -N_2 \frac{d(BA)}{dt} \quad \text{and} \quad V_1 = -N_1 \frac{d(BA)}{dt}$$

7.7. MUTUAL INDUCTANCE

These two equations lead to the following relation

$$\frac{V_1}{V_2} = \frac{N_1}{N_2}. \qquad (7.28)$$

This expression is referred to as the *transformer equation*. A transformer is a device that takes advantage of Faraday's law to step-up or step-down AC voltages. These devices are used widely throughout our electrified society.

To get the fields to nearly perfectly couple, as we have assumed in this derivation, the insulated wire is wound around an iron core. Iron is a very effective conductor of magnetic fields. A cartoon-like depiction of an iron core transformer is shown in Figure 7.12.

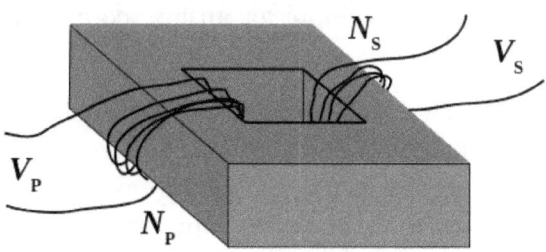

Figure 7.12: Drawing of an iron core transformer with a primary and secondary side. The body of the transformer is a solid block of iron with an inner hole. The iron acts to almost perfectly couple the field from one coil to the other.

Typically, when dealing with transformers one selects one voltage, or side, to be the primary, the other the secondary. Letting the primary side voltage be given by V_P

and primary side number of turn as N_P, and using an analogous notation for the secondary, leads to the transformer equation being written as

$$\frac{V_P}{V_S} = \frac{N_P}{N_S} \ . \tag{7.29}$$

The circuit symbol for an iron core transformer is given in Figure 7.13.

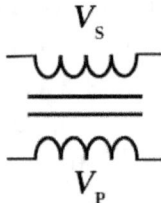

Figure 7.13: Circuit symbol for an iron core transformer.

Since the device can be used to "step-up" or increase a voltage, one might be tempted to think that it creates energy. However, this is not so. Energy is conserved when using the transformer, in fact it must be that the power at the primary equals the power at the secondary or

$$I_P V_P = I_S V_S \ . \tag{7.30}$$

Since transformers are used within AC circuits we will assume that the voltages here are *rms* voltages.

Example 7.6

Power companies routinely step up AC voltages at power stations before transmission of the electricity over long distances.

7.7. MUTUAL INDUCTANCE

This stepped up voltage reduces the i squared r losses associated with transmission cables.

a) A power provider wants to step up 400 V rms to 20 kV rms before transmission over a triple naught aluminum wire. If the number of turns on the primary side of the step-up transformer are 100, how many are required on the secondary?

b) Assume the distance of transmission is 20.0 km. Compare the i squared r losses for the case where the voltage remains at 400 V with the situation where it is stepped up to 20 kV.

a) Here we simply solve Eq. (7.28) for N_S and insert numbers

$$N_S = N_P \frac{V_S}{V_P} = (100) \frac{20,000}{400} = 5,000 \quad \text{turns.}$$

b) To compare the power dissipation in the transmission cable at the two different voltages we need only consider the current in the wire for the two cases. The power lines serve a community with homes and businesses that can be thought of as a resistive load R_L. So, in either case, the power consumed by the load would be the same. Let the low voltage current be I_l and the high voltage current be I_h we must have that

$$I_h V_h = I_l V_l \quad \text{(i)},$$

The power lost in the cable in the low voltage case is

$$P_l = I_l^2 R \quad \text{(ii)},$$

where R is the resistance of the wire. Using Eq. (i) in Eq. (ii) yields

$$P_l = I_h^2 \left(\frac{V_h}{V_l}\right)^2 R \quad \text{(iii)}.$$

Now, the power lost in the cable in the high voltage case is

$$P_h = I_h{}^2 R \quad \text{(iv)} .$$

Using Eq. (iv) in Eq. (iii) we are lead to

$$P_l = P_h \left(\frac{V_h}{V_l}\right)^2 \quad \text{(v)} .$$

Inserting voltages into Eq. (v) we get

$$P_l = 2500 P_h .$$

So the power lost in the transmission line in delivering the load via low voltage is 2,500 times greater than that lost when serving the load with the stepped up voltage.

It is often convenient to have one primary coil along with multiple secondary coils on the same transformer. Such a transformer is often referred to as a *multi-load transformer*. A multi-load transformer with two secondary coils is schematically depicted in Figure 7.14.

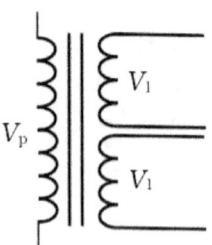

Figure 7.14: Schematic diagram of a multi-load iron core transformer with one primary and two secondary coils.

Assuming perfectly coupling of flux we arrive at

$$\frac{V_p}{N_p} = \frac{V_1}{N_1} = \frac{V_2}{N_2} . \qquad (7.31)$$

7.8. MAGNETIC CIRCUITS

As with the two coil transformer, the power of the primary is still equal to the power of the secondary so that

$$V_p I_p = V_1 I_1 + V_2 I_2 . \qquad (7.32)$$

7.8 Magnetic Circuits

With the idea of self-inductance now established, it is convenient to generalize our theory of the solenoid, or electromagnetic, to account for the effect of differing magnetic materials inside the core. We do this by developing a *magnetic circuit*. As the classical DC circuit deals with resistance and current, the analogues of these quantities in the magnetic circuit are *reluctance* and magnetic flux. Also, one might deal with the reciprocal of reluctance, *permeance*.

The permeance, P, of a rectangular magnetic sample of cross-sectional area A and length l is simply defined as

$$P = \frac{\mu A}{l} , \qquad (7.33)$$

where μ is the permittivity of the sample. As with inductance, permeance has the SI unit of Henry so that reluctance has the unit, inverse Henry.

Now, to deal with more than one type of magnetic material in a circuit, we consider the case where two rectangular samples, one of μ_1 the other μ_2 are arranged in a stack so that the internal field is directed along the length of the samples. Here we say the samples are arranged in parallel. In this case, the total permeance of the assembly, P is simply the sum of the permeances of the samples, that is

$$P = P_1 + P_2 . \qquad (7.34)$$

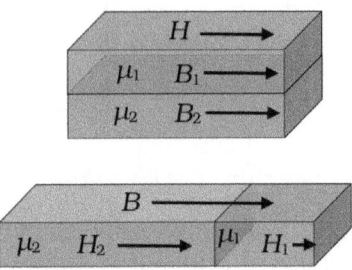

Figure 7.15: Sketch of two magnetic materials in parallel, top, and two in series, bottom.

However, if two different magnetic materials are placed end-to-end with the flux directed along their length we say that the samples are in series. In this case the reluctances are additive that is,

$$R_T = R_1 + R_2 \ . \tag{7.35}$$

These two situations are depicted in Figure 7.15. Note that since there is a discontinuity in the tangential component of a magnetic field at a boundary the magnetic field will be different in each material in the parallel arrangement while the H field is the same in both regions. The opposite is true for the series situation.

The main governing relation for the magnetic circuit is a sort of Ohm's law for magnetic circuits which relates the magnetic flux, Φ, through the material to the reluctance R. As the flux is typically created by a coil of n turns with current I we have that

$$nI = \Phi R \ . \tag{7.36}$$

7.8. MAGNETIC CIRCUITS

Here, one can think of nI as a sort of magnetic voltage, V_m, with units amp-turns.

The main result of this section is that we are able to use the ideas above to compute the inductance of a magnetic circuit that may be composed of several different materials. The final piece of the puzzle is found by considering again Faraday's law for an inductor: $\mathscr{E} = -d\Phi/dt$. In this case, Φ is a function of $I(t)$ so that we can write Faraday's law as

$$\mathscr{E} = -\frac{d\Phi}{dI}\frac{dI}{dt} \ . \tag{7.37}$$

On comparing this with Eq. (7.14), we conclude that

$$L = \frac{d\Phi}{dI} \ . \tag{7.38}$$

Example 7.7

An electromagnetic, depicted in Figure 7.16, is made of a rectangular iron core of cross-sectional area A and permeability μ with an air gap of width d. A current I is sent through the coil of n turns. Sketch a circuit diagram for the magnetic circuit description of the device and determine its inductance.

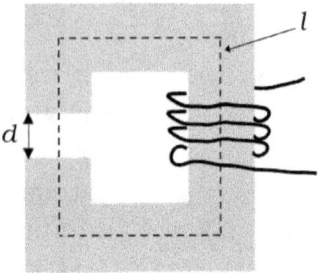

Figure 7.16: Sketch of an iron core electromagnet with an air gap.

The magnetic circuit is composed of a source, $V_m = nI$ and two reluctances, R_1 due to the metal core and R_2 due to the air gap. These relutances are in series therefore the magnetic circuit schematic is

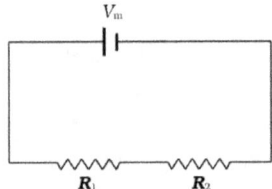

From the figure, the distance around our circuit is l, then from Eq. (7.33) the reluctance of the metal core R_1 will be

$$R_1 = \frac{l-d}{\mu A}.$$

The reluctance of the air gap is

$$R_1 = \frac{d}{\mu_o A}.$$

In series reluctances are additive so that the total reluctance R is

7.8. MAGNETIC CIRCUITS

$$R = \frac{1}{A}\left(\frac{(l-d)\mu_o + d\mu}{\mu\mu_o}\right).$$

Now, using Eq. (7.36) we get the magnetic flux Φ:

$$\Phi = \frac{nIA\mu\mu_o}{(l-d)\mu_o + d\mu}.$$

Finally, we use Eq. (7.38) to compute the inductance L:

$$L = \frac{nA\mu\mu_o}{(l-d)\mu_o + d\mu}.$$

PROBLEMS

7.9 A time varying, uniform magnetic field is given by $\vec{B}(t) = 2.2t\,\hat{x} - 1.75t\,\hat{z}$ T. What is emf induced through a single conducting loop that inscribes a square area of side length 20.0 cm which lies in the xy plane?

7.10 A magnetic field in spherical coordinates is $\vec{B} = B_o r\,\hat{r}$, where B_o is a constant. What is the flux of this field through a hemisphere of radius R whose open bottom is on the xy plane?

7.11 A conducting loop of one turn experiences a change in flux through its inscribed area of 0.23 Wb over a time period of 6.0 ms. The loop is connected to a load resistance of 200 Ω. What is the induced current through the loop?

7.12 A conducting rod of length 10.0 cm moves at 2.0 m/s along a u-shaped conductor in a uniform magnetic field of 0.28 T as shown in Fig. 16.6. Suppose that the system has a resistance of 1.2 Ω. What is the current induced in the loop?

7.13 A conducting loop rests in a spatially uniform, but time varying, magnetic field as shown in Fig. 16.28. If the magnitude of the magnetic field vectors are increasing in which direction is current induced in the loop?

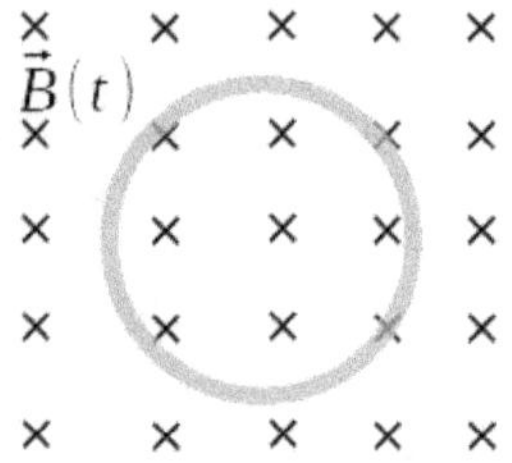

Figure 7.17: A conducting loop rests in a uniform time varying magnetic field.

7.14 A metal rail is slid along a u-shaped conductor with width 0.55 m as depicted in Figure 7.4. If the uniform magnetic field has magnitude 5.0 T, what force is required to maintain a speed of 10.0 m/s?

7.15 An inductor is required that has an inductance of 1.0 mH. You have a cylinder of circular cross section of radius 0.2 cm and length 1.2 cm. What minimum length of insulated wire is required?

7.16 It is easy to see that Eq. (7.22) can be generalized to yield the total energy, U, stored in a magnetic field over some region of space (say, all space), as

$$U = \frac{1}{2\mu_o} \int B^2 d\tau .$$

7.8. MAGNETIC CIRCUITS

A magnetized sphere of radius R emits a magnetic field $\vec{B} = \frac{B_o}{r^2}\hat{r}$ in the region $R \leq r < \infty$. Here B_o is a constant. Find the energy stored in this field in the region $R \leq r < \infty$.

7.17 Consider the magnetized upright cylinder from Example 6.1. If the cylinder length is l and the radius R, find the total energy stored in the magnetic field within the cylinder.

7.18 A coil of length l_1 of turns n_1 is wrapped around a non-magnetic cylindrical core. Another shorter coil of insulated wire of length l_2 and turns n_2 is wrapped over the longer coil as shown in the figure below. Find the mutual inductance of this arrangement.

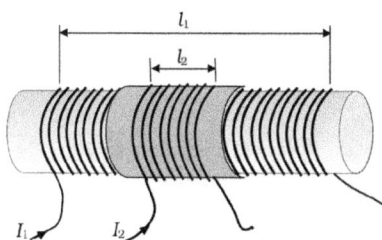

Figure 7.18: A primary coil with a secondary coil.

7.19 A power company uses a transformer to step up 220 V rms to 20 kV rms before transmission to a load of 5000.0 Ω. The secondary side of the transformer has a winding of 2500 turns.

a) How many turns for the winding of the primary are required?

b) What is the average power drawn by the load?

c) What is the rms current in the primary side of the transformer?

7.20 A transformer with one primary coil of 200 turns is connect to 120 V rms. One of the secondary coils must serve as a 25.0 V rms source while the other must be at 9.0 V rms. The 9.0 V secondary serves a load of 10.0 Ω. If the transformer can safely handle no more than 100.0 W,

a) what is the maximum current allowed through the other secondary coil?

b) what is the current at the primary coil?

c) what are the number of turns in each secondary coil?

7.21 Consider the electromagnetic depicted in the figure below.

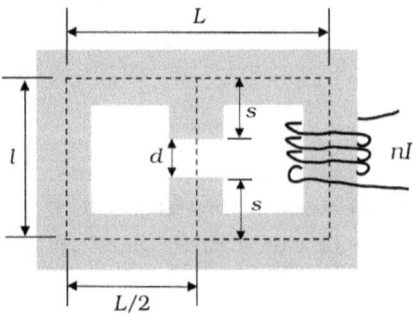

Figure 7.19: An electromagnet with an air gap.

7.8. MAGNETIC CIRCUITS

This electromagnet has an air gap of length d and has a constant cross-sectional area A. The core has permeability μ. Other relevant lengths are denoted in the figure.

a) Draw a schematic diagram for the magnetic circuit.

b) Find the inductance for the electromagnet.

c) Assuming a uniform field with no fringing, derive a formula for the magnetic field magnitude in the air gap in terms of nI and other relevant parameters.

7.22 Use Eq. (7.19), and the fact that current is common in a series circuit, to show that N inductors in series have an equivalent inductance, I_{eq}, given by

$$I_{eq} = I_1 + I_2 + I_3 + \cdots I_N .$$

7.23 Use Eq. (7.19), and the fact that voltage is common in a parallel circuit, to show that N inductors in parallel have an equivalent inductance, I_{eq}, given by

$$I_{eq} = \frac{1}{\frac{1}{I_1} + \frac{1}{I_2} + \frac{1}{I_3} + \cdots \frac{1}{I_N}} .$$

7.24 Consider the coaxial cable from Problem 5.33 where a DC current I travels down the central conductor and then back through the outer shield. Find the energy stored in the internal magnetic field in length l of the cable. Then, give the inductance per unit length.

Chapter 8

Maxwell's Equations

8.1 Displacement Current

By the mid 19th century, physicists had begun to realize that the electric and magnetic field were actually the same phenomenon. When viewed from a different frame of reference, an electric field is observed to be a magnetic field and vice versa.

The Scottish physicist James Clerk Maxwell (1831-1879), was able to show the unity of the electric and magnetic field in a elegant and quantitative way. He did this via a set of four equations that now bear his name, *Maxwell's equations*. Not only do these equations demonstrate the specific connection between electric and magnetic fields, they also provide additional information on the nature of these fields.

Even though this set of equations are named after Maxwell he was not the discoverer of all four. In fact, you have already encountered three of the four Maxwell equations at this point in your study. Our discussion of these four equations will begin by considering free space conditions with the possibility of charge being present.

The first of these is Gauss's law. Recall that Gauss's law states that the flux of an electric field through any closed surface is directly proportional to the enclosed charge, that is, as given by Eq. (2.18) or in differential form by Eq. (2.22). Adding to this idea, Maxwell hypothesized that there must be a Gauss's law for magnetic flux. In fact there is. However, in this case the flux of the magnetic field $\vec{B}$ through any closed surface area A is always zero:

$$\oint \vec{B} \cdot d\hat{A} = 0 . \tag{8.1}$$

What is implied by this fact? When the flux over a closed surface is zero we must have that the flux entering the surface always equals the flux leaving the surface. We would have this same result for a fluid flowing through a closed surface where there are no enclosed sinks or sources. However, in the case of a magnetic field we may have poles enclosed within the surface. The key issue is that for a magnetic field one will always have a north pole for every south pole. As mentioned previously, there are no known magnetic monopoles. This fact insures that the flux of the vector field through any closed surface will always be zero.

Taking the divergence of both sides of Eq. (8.1) it is easy to see that

$$\vec{\nabla} \cdot \vec{B} = 0 , \tag{8.2}$$

a result sometimes referred to as *Gauss's law for magnetic fields*.

8.1. DISPLACEMENT CURRENT

So two of the four Maxwell equations are Gauss's law for electric and magnetic fields. The next relation to be added to the list is Faraday's law. However, it is convenient to have it in a form explicitly involving the electric and magnetic field. To accomplish this, we use the integral formula for voltage from Eq. (3.13) in Faraday's law from Eq. (7.2) and arrive at

$$\oint \vec{E} \cdot d\vec{l} = -\frac{d\Phi}{dt} . \qquad (8.3)$$

Here we use the line integral around the closed path that inscribes the area through which Φ passes. Using the fundamental theorem for curls, the left side of Eq. (8.3) can be written as

$$\oint \vec{E} \cdot d\vec{l} = \int_S (\vec{\nabla} \times \vec{E}) \cdot d\vec{A} . \qquad (8.4)$$

Now, using this result, along with Eq. (7.1) for the flux in Eq. (8.3), we arrive at

$$\int_S (\vec{\nabla} \times \vec{E}) \cdot d\vec{A} = -\frac{\partial}{\partial t} \int_S \vec{B} \cdot d\vec{A} . \qquad (8.5)$$

Here, we have used partial differentiation as $\vec{B}$ may be a function of position as well as time. As the surface S is the same in both sides of Eq. (8.5) we conclude that

$$\vec{\nabla} \times \vec{E} = -\frac{\partial \vec{B}}{\partial t} , \qquad (8.6)$$

which is Faraday's law in differential form. So we have the first three of Maxwell's equations in differential form:

Gauss's law for electric fields, Eq. (2.22). Gauss's law for magnetic fields, Eq. (8.2) and Faraday's law given in Eq. (8.6). Now, we seek the differential form for the final Maxwell equation, Ampère's law.

Recall the integral form of Ampère's law from Eq. (5.46),

$$\oint \vec{B} \cdot d\vec{l} = \mu_o I , \qquad (8.7)$$

where we have let I be the enclosed current. Using the fundamental theorem for curls in Eq. (8.7) and using Eq. (5.3) to write I in terms of the current density J leads to

$$\int_S (\vec{\nabla} \times \vec{B}) \cdot d\vec{A} = \mu_o \int_S \vec{J} \cdot d\vec{A} . \qquad (8.8)$$

From this we conclude that

$$\vec{\nabla} \times \vec{B} = \mu_o \vec{J} . \qquad (8.9)$$

This is Ampère's law in differential form.

Though Eq. (8.9) is the final Maxwell equation it turns out that it is incomplete as written. Maxwell noticed that Ampère's law was incorrect for cases where there is a time varying charge density, ρ, present. To see this problem consider the divergence of both sides of Eq. (8.9).

$$\vec{\nabla} \cdot (\vec{\nabla} \times \vec{B}) = \mu_o (\vec{\nabla} \cdot \vec{J}) . \qquad (8.10)$$

Now, we know from our study of vector analysis that the divergence of a curl is zero. However, from the continuity equation, Eq. (5.21), the divergence of $\vec{J}$ is non-zero when $\partial \rho / \partial t$ is non-zero. To rectify this problem Maxwell

8.1. DISPLACEMENT CURRENT

proposed adding the term $\mu_o \partial \rho / \partial t$ to both sides of Eq. (8.9):

$$\vec{\nabla} \cdot (\vec{\nabla} \times \vec{B}) = \mu_o (\vec{\nabla} \cdot \vec{J}) + \mu_o \frac{\partial \rho}{\partial t} \ . \tag{8.11}$$

Using Gauss's law from Eq. (2.22) in this gives

$$\vec{\nabla} \cdot (\vec{\nabla} \times \vec{B}) = \mu_o (\vec{\nabla} \cdot \vec{J}) + \mu_o \epsilon_o \frac{\partial}{\partial t} (\vec{\nabla} \cdot \vec{E}) \ . \tag{8.12}$$

Factoring out the divergence in the above we are lead to the equality

$$\vec{\nabla} \times \vec{B} = \mu_o \vec{J} + \mu_o \epsilon_o \frac{\partial \vec{E}}{\partial t} \ . \tag{8.13}$$

This is the corrected form of Ampère's law and the fourth and final of Maxwell's equations. The last term on the right is often referred to as the *displacement current*.

All of Maxwell's equations for free space with charge can now be listed. These are encapsulated below. The student should commit these to memory.

$$\vec{\nabla} \cdot \vec{E} = \frac{\rho}{\epsilon_o} \qquad (8.14)$$

$$\vec{\nabla} \cdot \vec{B} = 0 \qquad (8.15)$$

$$\vec{\nabla} \times \vec{E} = -\frac{\partial \vec{B}}{\partial t} \qquad (8.16)$$

$$\vec{\nabla} \times \vec{B} = \mu_o \vec{J} + \mu_o \epsilon_o \frac{\partial \vec{E}}{\partial t} \qquad (8.17)$$

PROBLEMS

8.1 A time varying charge density in one dimension is given by $\rho = \rho_o \sin(\omega t)$ where ρ_o and ω are constants. Use Gauuss's law to find an expression for the electric field where $\vec{E} = E_o(x,t)\,\hat{x}$. Find the displacement current.

8.2 Use the fundamental theorem for curls, the definition for electric field flux, along with Eq. (5.21), to show that Ampère's law can be written in integral form as

$$\oint \vec{B} \cdot d\vec{l} = \mu_o I + \mu_o \epsilon_o \frac{\partial \Phi_E}{\partial t}\ .$$

8.3 Show that the divergence of Faraday's, (Eq. (8.16)), leads to Gauss's law for the magnetic field.

8.2 The Wave Equation

One of the fascinating results of the Maxwell equations is that they predict that the electric and magnetic fields will behave as traveling waves in charge free space. To see this, consider the curl of the left side of Eq. (8.16) along with the identity of Eq. (1.57):

$$\vec{\nabla} \times (\vec{\nabla} \times \vec{E}) = \vec{\nabla}(\vec{\nabla} \cdot \vec{E}) - \vec{\nabla} \cdot \vec{\nabla} \vec{E} . \quad (8.18)$$

Now in charge free space $\vec{\nabla} \cdot \vec{E} = 0$ and $\vec{\nabla} \cdot \vec{\nabla} \vec{E} = \nabla^2 \vec{E}$ so that the above becomes

$$\vec{\nabla} \times (\vec{\nabla} \times \vec{E}) = -\nabla^2 \vec{E} . \quad (8.19)$$

This result is then equated with the curl on the right side of Eq. (8.16) to give

$$\nabla^2 \vec{E} = \frac{\partial}{\partial t} \vec{\nabla} \times \vec{B} . \quad (8.20)$$

Using Eq. (8.17) to replace the curl of $\vec{B}$ in this for charge free space where $\vec{J} = 0$ gives

$$\nabla^2 \vec{E} = \mu_o \epsilon_o \frac{\partial^2 \vec{E}}{\partial t^2} . \quad (8.21)$$

In one spatial dimension this is simply

$$\frac{\partial^2 \vec{E}}{\partial x^2} = \mu_o \epsilon_o \frac{\partial^2 \vec{E}}{\partial t^2} . \quad (8.22)$$

One might recognize Eq. (8.22) as a wave equation. Therefore, Maxwell's equations predict that the electric field will be in the form of a traveling wave in charge free space. But there's more. If $\Psi = \Psi(x,t)$ is a traveling wave, then it obeys the classical wave equation

$$\frac{\partial^2 \Psi}{\partial x^2} = \frac{1}{v^2} \frac{\partial^2 \Psi}{\partial t^2}, \qquad (8.23)$$

where v is the wave speed. On comparing Eqs. (8.22) and (8.23) we find that the speed of the electric field traveling wave, which we will label as c, is given by

$$c = \frac{1}{\sqrt{\mu_o \epsilon_o}}. \qquad (8.24)$$

This result gives the well known speed of light in terms of physical constants that had been known some time before Maxwell assembled his equations. The fact that Eq. (8.24) correctly yields the experimentally known value for the speed of light in vacuum, 2.9×10^8 m/s, is truly remarkable.

Additional information about how $\vec{E}$ and $\vec{B}$ relate to one another can be found by considering a solution for Eq. (8.22) such as

$$\vec{E}(x,t) = E_o e^{i(kx - wt)} \hat{z}. \qquad (8.25)$$

Eq. (8.25) describes an oscillating wave of amplitude E_o which lies in the xz plane and propagates in the $+x$ direction as time t passes. We call such a wave a *plane polarized wave*. Here ω is the frequency and k the wave vector. The wave vector is related to the wavelength λ by

$$k = \frac{2\pi}{\lambda}. \qquad (8.26)$$

8.2. THE WAVE EQUATION

Using Eq. (8.25) in the left side of Eq. (8.16) and taking the curl yields

$$ikE_o e^{i(kx-wt)} \hat{y} = -\frac{\partial \vec{B}}{\partial t}. \tag{8.27}$$

This is true for the case where $\vec{B} = B_o e^{i(kx-wt)} \hat{y}$. Using this for $\vec{B}$ in the above gives

$$ikE_o e^{i(kx-wt)} \hat{y} = iwB_o e^{i(kx-wt)} \hat{y}. \tag{8.28}$$

So it must be that $kE_o = \omega B_o$ or $E_0/B_o = \omega/k$. The wave frequency in cycles per second f is related to the wavelength by

$$c = f\lambda. \tag{8.29}$$

Since $\omega = 2\pi f$ we conclude that

$$\frac{E_o}{B_o} = c. \tag{8.30}$$

So we see that Maxwell's equations predict that both the electric and magnetic fields are wave like in charge free space. Their amplitudes are related by Eq. (8.30) and when the electric field is plane polarized in the xz the corresponding magnetic field plane wave is polarized in the xy plane. That is, the planes of polarization are 90° apart. Such traveling waves are referred to as *electromagnetic*, (E&M), *radiation*.

A simplistic plot of the E&M traveling wave discussed above is given in Figure 8.1

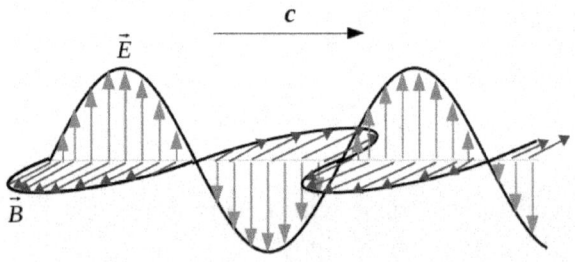

Figure 8.1: A plane polarized E&M traveling wave.

Example 8.1

Show that $\Psi(x,t) = A\sin(kx - \omega t)$ is a solution to the wave equation and that $v = \lambda f$.

Inserting this into the left side of Eq. (8.23) and differentiating twice gives

$$Ak^2 \sin(kx - \omega t) .$$

Now, we put Ψ into the right side of Eq. (8.23) and get

$$\tfrac{A\omega^2}{v^2} \sin(kx - \omega t) .$$

So we see that Ψ is a solution for Eq. (8.23) when $k^2 = \tfrac{\omega^2}{v^2}$. Since $k = 2\pi/\lambda$ this leads to

$$(\tfrac{2\pi}{\lambda})^2 = \tfrac{\omega^2}{v^2} .$$

Solving for v gives

8.2. THE WAVE EQUATION

$$v = \frac{\omega\lambda}{2\pi} = f\lambda .$$

PROBLEMS

8.4 Show that $\Psi = Ae^{-i(kx-\omega t)}$ is a solution for the wave equation and that $v = f\lambda$.

8.5 Using the approach taken in this section to arrive at Eq. (8.22), derive an analogous wave equation for $\vec{B}$.

8.6 The magnetic field component of an E & M wave is given by:

$$\vec{B} = (6.0 \times 10^{-4})\cos\left[(2.0 \times 10^8)t + 0.69y\right]\ \hat{z}\ \text{T}.$$

Determine the frequency, wavelength and electric field amplitude for this wave. Write the expression for the electric field component.

8.7 Show that the linear combination

$$\Psi = \sum_{n=1}^{\infty} A_n \sin(nkx - \omega t) ,$$

is a solution for the wave equation. Here A_n are constant coefficients and n is a positive integer.

8.3 Energy in E & M Waves

In previous chapters, we learned that electric and magnetic fields contain stored or, potential energy. Using the result for the energy stored in a capacitor, the energy density, u_o, for a static electric field was found to be $(1/2)\epsilon_o E^2$ and for a static magnetic field $(1/2)B^2/\mu_o$. Using these results we can write that the energy density in an E & M wave at some instant in time and space is given by

$$u_o = \frac{1}{2}\epsilon_o E^2 + \frac{1}{2}\frac{B^2}{\mu_o} \ . \qquad (8.31)$$

where here E and B represent the instantaneous magnitude of the electric and magnetic field respectively. Using Eqs. (8.24) and (8.30) the above can be written as

$$u_o = \epsilon_o E^2 \ . \qquad (8.32)$$

A more interesting energy related quantity, for an E & M wave, is the energy passing through unit area A per unit time. Suppose the wave goes a distance cdt in time dt. Then we define a differential of volume dV as $dV = Acdt$. Therefore, the differential for energy, U, is then

$$dU = u_o A c dt = \epsilon_o E^2 A c dt \ . \qquad (8.33)$$

Now, we denote the energy passing through unit area per unit time as S. Using the above result this is

$$S = \frac{1}{A}\frac{dU}{dt} = c\epsilon_o E^2 \ . \qquad (8.34)$$

You may recall from our work on sound that an energy per unit area per unit time is referred to as an intensity

8.3. ENERGY IN E & M WAVES

and has the SI unit W/m². As with Eq. (8.31), Eq. (8.34) gives the wave intensity at a particular point in space and time. A more useful expression for S is obtained by reinserting the magnetic field magnitude into this result. Using Eqs. (8.24) and (8.30) we arrive at

$$S = \frac{1}{\mu_o} EB. \qquad (8.35)$$

The above result now motivates the creation of a vector, $\vec{S}$, whose magnitude has units of intensity. We call $\vec{S}$ the *Poynting vector* named after J. H. Poynting (1852-1914).

$$\vec{S} = \frac{1}{\mu_o}(\vec{E} \times \vec{B}). \qquad (8.36)$$

Ironically, $\vec{S}$ "points" in the direction of propagation.

Eq. (8.35) gives the instantaneous magnitude for $\vec{S}$. We can write an expression for the mean value of the magnitude of $\vec{S}$, $\langle S \rangle$, in terms of the amplitudes of B and E, which we denote as B_o and E_o respectively,

$$\langle S \rangle = \frac{E_o B_o}{2\mu_o}. \qquad (8.37)$$

The intensity of a plane wave, as in the cases considered above, would remain constant as it travels. However, radiation typically comes from a point-like source and then spreads out over an ever greater surface area as it propagates away. Suppose we have a point source for E & M radiation that emits energy uniformly in all directions. That is, there is spherical symmetry about the source. We can then model the intensity of the source versus distance using the classic inverse square relationship. Assuming that the power output, P, of the source is constant, and with our scalar E & M intensity given by S, we can then write,

$$S = \frac{P}{4\pi r^2}, \qquad (8.38)$$

where r is the radial distance from the source. We can state the above more generally as

$$P = \int \vec{S} \cdot d\vec{A}. \qquad (8.39)$$

8.4 Maxwell's Equations in Materials

In the previous section Maxwell's equations were derived for free space with free charge, free space with no charge and even free space with a free current density. In this section we want to consider these equations for situations where we have a dielectric and/or magnetic material along with the possibility of free charge. Recall that when there is an electric field within a dielectric material, there is a possibility of there being some bound charge density ρ_b. Therefore we have to first adjust Gauss's law for this fact as follows:

$$\vec{\nabla} \cdot \vec{E} = \frac{\rho_f + \rho_b}{\epsilon_o}. \qquad (8.40)$$

Using Eq. (4.8) we can write ρ_b in terms of the polarization $\vec{P}$ to get

$$\vec{\nabla} \cdot \vec{E} = \frac{\rho_f - \vec{\nabla} \cdot \vec{P}}{\epsilon_o}. \qquad (8.41)$$

Re-arranging this result, and using the fact that the displacement field $\vec{D}$ is defined as $\vec{D} = \epsilon_o \vec{E} + \vec{P}$, we are lead to

$$\vec{\nabla} \cdot \vec{D} = \rho_f. \qquad (8.42)$$

8.4. MAXWELL'S EQUATIONS IN MATERIALS

Now, Gauss's law for magnetic fields and Faraday's law both require no changes as they do not explicitly involve charge or current. However, Ampère's law will require a little work. One might think that it is a simple process of replacing $\vec{J}$ with $\vec{J}_f + \vec{J}_b$ in Eq. (8.17) but this is not the case. It turns out that there is another type of current density that can be produced when there is a time varying electric field present in a dielectric. This is referred to as the *polarization current*, $\vec{J}_p$. Given the name of this current density, it should be a surprise that

$$\vec{J}_p = \frac{\partial \vec{P}}{\partial t} . \tag{8.43}$$

So, for our new version of Ampère's law we need to consider a sum of three current densities. That is,

$$\vec{J}_f + \vec{J}_p + \vec{J}_b = \vec{J}_f + \frac{\partial \vec{P}}{\partial t} + \vec{\nabla} \times \vec{M} . \tag{8.44}$$

where we have used Eq. (6.1) to relate $\vec{J}_b$ to the magnetization $\vec{M}$. Using Eq. (8.44) in Eq. (8.17) yields

$$\vec{\nabla} \times \vec{B} = \mu_o \left(\vec{J}_f + \frac{\partial \vec{P}}{\partial t} + \vec{\nabla} \times \vec{M} \right) + \mu_o \epsilon_o \frac{\partial \vec{E}}{\partial t} . \tag{8.45}$$

Subtracting $\vec{\nabla} \times \vec{M}$ from both sides of this, then using Eq. (4.17) to replace $\vec{E}$ and $\vec{P}$ with $\vec{D}$ and Eq. (6.5) to replace $\vec{B}$ with $\vec{H}$ we get the final result:

$$\vec{\nabla} \times \vec{H} = \vec{J}_f + \frac{\partial \vec{D}}{\partial t} . \tag{8.46}$$

This concludes the explanation of Maxwell's equations in dielectric and magnetic materials for the cases where there may be free charges and free currents. These are all listed together below.

$$\vec{\nabla} \cdot \vec{D} = \rho_f \qquad (8.47)$$

$$\vec{\nabla} \cdot \vec{B} = 0 \qquad (8.48)$$

$$\vec{\nabla} \times \vec{E} = -\frac{\partial \vec{B}}{\partial t} \qquad (8.49)$$

$$\vec{\nabla} \times \vec{H} = \vec{J}_f + \frac{\partial \vec{D}}{\partial t} \qquad (8.50)$$

Example 8.2

In some material it is known that $\vec{H} = H_o(y\,\hat{z} - x\,\hat{y})$ and $\vec{J}_f = J_o\,\hat{x}$ where H_o and J_o are constants and all units are SI. Determine $\vec{D}$ with the initial condition $\vec{D} = 0$ at $t = 0$.

Use Ampère's law in materials, Eq. (8.50). Computing the curl of $\vec{H}$ gives:

$$\vec{\nabla} \times \vec{H} = -2H_o\,\hat{z} \ .$$

We write $\vec{D}$ in terms of unknown Cartesian components and then insert everything into Eq. (8.50):

$$-2H_o\,\hat{z} = J_o\,\hat{x} + \tfrac{d}{dt}(D_x\,\hat{x} + D_y\,\hat{y} + D_z\,\hat{z}) \ .$$

Now, equate like coefficients: $\frac{dD_x}{dt} = -J_o$.
Separate and integrate to get

$$D_x = -J_o t + A, \text{ where } A \text{ is a constant.}$$

Also, $\frac{dD_y}{dt} = 0$. So, it must be that $D_y = B$ where B is a constant.

Finally $\frac{dD_z}{dt} = -2H_o$.
Separate and integrate to get

$$D_z = -2H_o t + C, \text{ where } C \text{ is a constant.}$$

Putting it all together

$$\vec{D} = (-J_o t + A)\,\hat{x} + B\,\hat{y} + (-2H_o t + C)\,\hat{z}.$$

Since our initial condition is $\vec{D} = 0$ at $t = 0$, all of the constants must equal zero and our particular displacement field is

$$\vec{D} = -J_o t\,\hat{x} - 2H_o t\,\hat{z}.$$

PROBLEMS

8.8 The infrared (IR) band of the E & M spectrum is often divided into sub-bands. Two of these are the near-infrared (NIR) and the far-infrared (FIR). The wavelength range for the NIR is 0.75-1.4 μm. The range for the FIR is 15-1000 μm. IR Spectroscopists routinely report IR wavelengths (frequencies) as *wave numbers*, where this wave number is defined as $1/\lambda$, where λ is in cm. Therefore, these values for wave numbers are in units of cm^{-1}, (inverse centimeters). Compute the corresponding spectroscopic wave number range for each of these bands. Give the wave number units in cm^{-1}.

8.9 An E & M wave has as its electric field component

$$\vec{E} = (0.12)\cos\left[(10.4 \times 10^6)y - \omega t\right] \hat{z} \text{ V/m}.$$

Write the expression for the magnetic field component of this wave and determine the Poynting vector and the average intensity.

8.10 Use the appropriate integral theorem to write Eqs. (8.47) – (8.50) in integral form. (There should be no current or charge densities in your final result.)

8.5 Maxwell's Equations and Conductors

A special situation involving Maxwell's equations is when we have a linear material present which is also a conductor. Though the conductor is uncharged, it still contains charge that is free to move so that there can be a free current. As it happens, a peculiar phenomenon will occur in this case and it is instructive to study it further.

In this situation Gauss's law for electric fields becomes $\vec{\nabla} \cdot \vec{E} = 0$. Since we want to study the behavior of the electric field, we will use Gauss's law along with a modified version of Eq. (8.17) to arrive at the result. Since we have a conductor, then by Ohm's law, $\vec{J} = \sigma \vec{E}$. Using

8.5. MAXWELL'S EQUATIONS AND CONDUCTORS

this in Eq. (8.17) gives

$$\vec{\nabla} \times \vec{B} = \mu \sigma \vec{E} + \mu \epsilon \frac{\partial \vec{E}}{\partial t} . \tag{8.51}$$

Taking the time derivative of both sides of this yields

$$\vec{\nabla} \times \frac{\partial \vec{B}}{\partial t} = \mu \sigma \frac{\partial \vec{E}}{\partial t} + \mu \epsilon \frac{\partial^2 \vec{E}}{\partial t^2} . \tag{8.52}$$

Using Faraday's law (Eq. (8.16)) to write $\partial \vec{B}/\partial t$ in terms of $\vec{E}$ leads to

$$\vec{\nabla} \times (-\vec{\nabla} \times \vec{E}) = \mu \sigma \frac{\partial \vec{E}}{\partial t} + \mu \epsilon \frac{\partial^2 \vec{E}}{\partial t^2} . \tag{8.53}$$

Expanding the triple product according to Eq. (1.38), using the fact that $\vec{\nabla} \cdot \vec{E} = 0$ here, and re-arranging yields the final governing equation:

$$\nabla^2 \vec{E} = \mu \epsilon \frac{\partial^2 \vec{E}}{\partial t^2} + \mu \sigma \frac{\partial \vec{E}}{\partial t} . \tag{8.54}$$

Here we will consider this result only in one spatial dimension say x. Notice how Eq. (8.54) is just the classical wave equation (Eq. (8.22)) with a damping term.

Eq. (8.54) can be solved by the method of separation of variables but we might make a guess that the traveling wave is a solution. This turns out to be the case and gets us more directly to the desired result. So, let the solution to Eq. (8.54) be

$$\vec{E}(x,t) = E_o e^{i(kx - \omega t)} . \tag{8.55}$$

Inserting this into Eq. (8.54) and performing the required differentiation leads to

$$k = \sqrt{\mu \epsilon \omega^2 + i \, \mu \sigma \omega} , \tag{8.56}$$

Unlike the case of the undamped wave equation, here our wave number has an imaginary component. It is useful to break k into real and imaginary terms. However, the precise nature of these terms is obscured by the radical in Eq. (8.56). We can find these by considering the following. We seek constants A and B such that (B is not a magnetic field magnitude),

$$(A + i\,B)^2 = \mu\epsilon\omega^2 + i\,\mu\sigma\omega , \qquad (8.57)$$

Expanding the left side of this and equating coefficients gives

$$-2AB = \mu\sigma\omega \quad \text{and} \quad A^2 - B^2 = \mu\epsilon\omega^2 .$$

Using these to eliminate B gives the following for A:

$$4A^4 - 4\mu\epsilon\omega^2 A^2 - (\mu\sigma\omega)^2 = 0 . \qquad (8.58)$$

This can be solved to yield

$$A^2 = \frac{\mu\epsilon\omega^2}{2} + \frac{\sqrt{(\mu\sigma\omega^2)^2 + (\mu\epsilon\omega)^2}}{2} . \qquad (8.59)$$

Taking the square root of both sides of this, and doing a bit of simplification, gives the final result for A:

$$A = \omega\sqrt{\frac{\mu\epsilon}{2}} \left[1 + \sqrt{1 + \left(\frac{\sigma}{\epsilon\omega}\right)^2} \right]^{1/2} . \qquad (8.60)$$

Now, using some of the relations above we find that

$$B = \omega\sqrt{\frac{\mu\epsilon}{2}} \left[-1 + \sqrt{1 + \left(\frac{\sigma}{\epsilon\omega}\right)^2} \right]^{1/2} . \qquad (8.61)$$

8.5. MAXWELL'S EQUATIONS AND CONDUCTORS

Now we have the wave number in the appropriate form, that is

$$k = A + i\,B\ . \tag{8.62}$$

Inserting this into Eq. (8.55) gives

$$\vec{E} = E_o e^{-Bx} e^{i(Ax-\omega t)}\ . \tag{8.63}$$

We recognize this result as a plane wave with a decaying exponential envelope function. The amplitude of the wave has decayed by about 37% at the distance

$$\delta = \frac{1}{B}\ , \tag{8.64}$$

where δ is called the *skin depth*. This result helps to describe why some materials are transparent at certain frequencies while others are opaque.

Example 8.3

Compute the skin depth for copper.

Use the resistivity for copper given in Table 5.1. Let $\mu = \mu_o$ and $\epsilon = \epsilon_o$. We select the frequency from approximately the middle of the visible spectrum, say 550 nm. Therefore, our conductivity is 5.9×10^7 $(\Omega m)^{-1}$ and $\omega = 3.3 \times 10^{15}$ rad/s. Putting all of this into Eq. (8.61) and then reciprocating gives:

$$2.86 \text{ nm}\ .$$

So the wave only penetrates 30 or so angstroms so that the metal would appear opaque in the visible as expected.

8.6 Electric Dipole Radiation

In the previous sections we have studied the wave nature of E&M radiation but we have not said much about the origin of these waves. The source of E&M radiation is accelerating charge. More practically, an alternating current should then generate, or radiate, E&M radiation. In this section we will study the simplest of such radiators, the *electric dipole radiator*. This is a simple model that can be used to describe the nature of radiation emitted by most broadcasting antennae.

To create a simple dipole radiator, an AC source is connected to two conducting wires arranged as shown in Figure 8.2

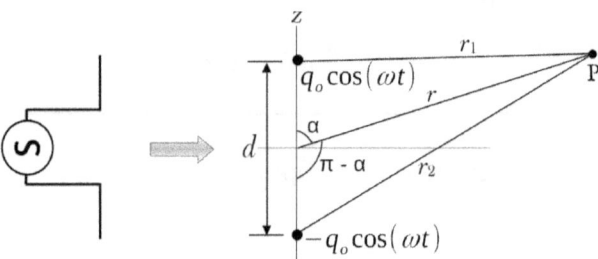

Figure 8.2: A dipole radiator. Source and radiating antenna on the left. Two oscillating point charge model on the right.

Our goal here will be to get the Poynting vector for this radiator. With this, an expression for the average intensity as a function of position can be determined. However, to get the Poynting vector we need $\vec{E}$ and $\vec{B}$ and to get these we will need some potentials.

8.6. ELECTRIC DIPOLE RADIATION

You may recall from working on Problem 5.32, that since $\vec{\nabla} \cdot \vec{B} = 0$ one can write $\vec{B}$ in terms of a vector potential $\vec{A}$ as

$$\vec{B} = \vec{\nabla} \times \vec{A}. \tag{8.65}$$

This is true for static and time varying magnetic fields.

Also recall that for static electric fields where $\vec{\nabla} \times \vec{E} = 0$, $\vec{E}$ can be written in terms of some scalar potential V as $\vec{E} = -\vec{\nabla} V$. However, for time varying electric fields this is not the case. We can adjust for this by writing $\vec{E}$ in terms of a scalar potential and $\vec{A}$ using Faraday's law. Using Eq. (8.65) in Eq. (8.16) gives

$$\vec{\nabla} \times \vec{E} = -\frac{\partial}{\partial t}(\vec{\nabla} \times \vec{A}). \tag{8.66}$$

Upon re-arranging it is clear that

$$\vec{\nabla} \times \left(\vec{E} + \frac{\partial \vec{A}}{\partial t} \right) = 0. \tag{8.67}$$

So it must be that the stuff in parenthesis of Eq. (8.67) can be written in terms of a scalar potential, say V, and thus

$$\vec{E} = -\frac{\partial \vec{A}}{\partial t} - \vec{\nabla} V. \tag{8.68}$$

Therefore, given the potentials V and $\vec{A}$, we can find the fields $\vec{E}$ and $\vec{B}$, and thus $\vec{S}$, for our dipole oscillator.

In this analysis we will make several approximations that will greatly simplify the analysis but still leave us with a useful result. First we will assume when a signal leaves a point charge in the dipole it is immediately at the observation point P. That is, we ignore any effects due to a phenomenon referred to as *retardation*. Also, we will assume that $r \gg d$ and if λ is the wavelength of the radiation then $r \gg \lambda$. With these restrictions in mind, it is a straight forward thing to use Eq. (3.20) to get an expression for the scalar potential V:

$$V = \frac{1}{4\pi\epsilon_o} \left[\frac{q_o \cos(\omega t)}{r_1} - \frac{q_o \cos(\omega t)}{r_2} \right] . \qquad (8.69)$$

But, rather than deal with the drudgery required to manipulate this expression for the voltage, we might just propose that the voltage in this case would be just that of a static dipole, as from Problem 4.33, multiplied by a time dependent factor such as

$$V = \frac{p \cos\theta}{4\pi\epsilon_o r^2} \cos(\omega t) , \qquad (8.70)$$

where $p = q_o d$. It turns out, this would be a good guess.

Now, for large distances from the source we would expect that V would vary as $1/r$ as predicted by Coulomb's law. If we are considering an oscillator in the radio region of the spectrum then its a good approximation to let $1/r^2 \approx \omega/(rc)$ for large $r \gg d$. Also, this substitution keeps the units straight. This final approximation gives us the final expression for the voltage:

$$V = \frac{p\omega \cos\theta}{4\pi\epsilon_o c \, r} \cos(\omega t) . \qquad (8.71)$$

With V in hand we now deal with getting an expression for $\vec{A}$. Considering how the voltage relates to a charge density in Eq. (3.21), we then, through an analogous expression, relate the vector potential, a sort of magnetic field voltage, to a current density as

$$\vec{A} = \frac{\mu_o}{4\pi} \int \frac{\vec{J}}{r} d\tau . \qquad (8.72)$$

8.6. ELECTRIC DIPOLE RADIATION

Here, we just have a wire with a line current in it so that this integral is reduced to

$$\vec{A} = \frac{\mu_o}{4\pi} \int_{-d/2}^{d/2} \frac{(dq/dt)\,\hat{z}}{r} dz \;. \tag{8.73}$$

Taking $q(t)$ from Figure 8.2 this integral can be computed to give

$$\vec{A} = -\frac{\mu_o p \omega}{4\pi r} \sin(\omega t)\,\hat{z} \;. \tag{8.74}$$

Now we are in a position to compute $\vec{E}$ and $\vec{B}$ for our radiator. Since we have the radial coordinate r, we use spherical coordinates. We get $\vec{E}$ from (8.68). First we can simplify the outcome by realizing that the terms in $-\vec{\nabla}V$ will all be of order $1/r^2$ and can be neglected for large r. This leaves just $-\partial\vec{A}/\partial t$ to get $\vec{E}$. Using Eq. (8.74) in this gives

$$\vec{E} = -\frac{\mu_o p \omega^2}{4\pi r} \cos(\omega t) \left(\cos\theta\,\hat{r} - \sin\theta\,\hat{\theta} \right) \;, \tag{8.75}$$

where we have used $\hat{k}$ in spherical coordinates from an Appendix. We can further simplify this by just letting our $\vec{E}$ be that of a traveling spherical wave. That is, we will ignore the $\hat{r}$ component of $\vec{E}$. Additionally, this will make the Poynting vector easier to compute. So our final expression for $\vec{E}$ is

$$\vec{E} = -\frac{\mu_o p \omega^2}{4\pi r} \sin\theta \cos(\omega t)\,\hat{\theta} \;. \tag{8.76}$$

Get $\vec{B}$ from the curl of $\vec{A}$. This gives

$$\vec{B} = -\frac{\mu_o p \omega}{4\pi r^2} \sin\theta \sin(\omega t)\,\hat{\phi} \;. \tag{8.77}$$

In this we will use our approximation $1/r^2 \approx \omega/(rc)$ again. Also, the $\vec{B}$ field may as well be set in phase with the $\vec{E}$ field. This gives

$$\vec{B} = -\frac{\mu_o p \omega^2}{4\pi cr} \sin\theta \cos(\omega t) \, \hat{\phi} \, . \qquad (8.78)$$

Now, it is a straight forward thing to use Eq. (8.76) and Eq. (8.78) to compute the Poynting vector for the oscillating dipole:

$$\vec{S} = \frac{1}{\mu_o} = \frac{\mu_o}{c} \left[\frac{p\omega^2}{4\pi r} \sin\theta \cos(\omega t) \right]^2 \hat{r} \, . \qquad (8.79)$$

It will be left as a problem to show how Eq. (8.79) can be used to obtain the average intensity of the radiator and average total power radiated.

PROBLEMS

8.11 Use the appropriate integral theorems to write Eqs. (8.14) – (8.16) in integral form.

8.12 A time dependent electric field is given by $\vec{E} = E_o(yt^p \, \hat{x} + zt^p \, \hat{y})$ where E_o is a constant and p is an integer. Use Faraday's law to find the corresponding magnetic field.

8.13 In one dimensional space a charge density is $\rho = \rho_o e^{-kx}$, where ρ_o and k are constants. Determine the electric field.

8.6. ELECTRIC DIPOLE RADIATION

8.14 An E & M wave has as its electric field component

$$\vec{E} = (0.12)\cos\left[(10.4 \times 10^6)y - \omega t\right] \hat{z} \text{ V/m}.$$

What is the energy density for the wave at $y = 10$ cm and $t = 2.0$ s?

8.15 The Sun delivers E & M radiation of an intensity of approximately 1000 W/m² on the Earth. Astronomers tell us that the Sun is around 1.5×10^{11} m from Earth. Use this distance to estimate the power output of the Sun. Repeat the calculation for an Earth-Sun distance of 2,000 miles.

8.16 The resistivity for common glass is around 1.0×10^{11} Ωm. Compute the skin depth for glass in the visible. What does this result predict concerning the appearance of glass?

8.17 Estimate the skin depth for visible light in the materials listed in Table 5.1.

8.18 The electric field component of a plane E&M wave is given by $\vec{E} = E_o e^{-i(kx-\omega t)} \hat{y}$ where E_o is the wave amplitude. Find the corresponding magnetic field wave and then compute the Poynting vector for this wave. Show that the average intensity fof this wave is given by Eq. (8.37).

8.19 Take the divergence of both sides of Eq. (8.68) and show that $\nabla^2 V + \frac{\partial}{\partial t}(\vec{\nabla} \cdot \vec{A}) = -\rho/\epsilon_o$.

8.20 It is convenient in static situations to let $\vec{\nabla} \cdot \vec{A} = 0$. Using this convention, along with Eqs. (8.65) and (8.68) in the differential form of Ampère's law, show that one can get the following differential equation for $\vec{A}$:

$$\nabla^2 \vec{A} - \mu_0 \epsilon_0 \frac{\partial^2 \vec{A}}{\partial t^2} = \mu_0 \vec{J} + \mu_0 \epsilon_0 \vec{\nabla}\left(\frac{\partial V}{\partial t}\right) .$$

8.21 Show that the curl of Eq. (8.74) yields Eq. (8.77).

8.22 Use the mean value theorem from calculus, and Eq. (8.79), to find the average intensity of a dipole radiator over one cycle. Use this result in Eq. (8.39) to find the average power output of the dipole radiator.

8.23 For radiating antennae, a quantity called the *directivity*, D is of interest to know. It is a measure of the amount of the output power radiated in a particular direction. For the dipole radiator it can be approximately given by

$$D(\theta) = 1.64 \left[\frac{\cos\left(\frac{\pi}{2} \cos\theta\right)}{\sin\theta}\right]^2 .$$

At what angles θ is D a maximum? At what angles θ is D a minimum?

8.24 A static magnetic field is given by $\vec{B} = ax\,\hat{x} + \frac{b}{x}\,\hat{y}$ T, where a and b are unknowns. For what values of a and b does this field obey Maxwell's second equation. That is, one must have that $\vec{\nabla} \cdot \vec{B} = 0$.

Appendices

A Integral Formulas

$$\int e^{ax}\,dx = \frac{e^{ax}}{a} \tag{A.1}$$

$$\int xe^{ax}\,dx = \frac{e^{ax}}{a^2}(ax-1) \tag{A.2}$$

$$\int \frac{dx}{x} = \ln x \tag{A.3}$$

$$\int \frac{dx}{a+x} = \ln(a+x) \tag{A.4}$$

$$\int \frac{1}{a+bx}\,dx = \frac{1}{b}\ln|(a+bx)| \tag{A.5}$$

$$\int \sqrt{x^2 \pm a^2}\,dx = \frac{1}{2}\left[x\sqrt{x^2 \pm a^2} \pm a^2 \ln\left(x+\sqrt{x^2 \pm a^2}\right)\right] \tag{A.6}$$

$$\int \frac{1}{\sqrt{x^2 \pm a^2}}\,dx = \ln\left(x \pm \sqrt{x^2 \pm a^2}\right) \tag{A.7}$$

$$\int \frac{x}{\sqrt{x^2 \pm a^2}}\,dx = \sqrt{x^2 \pm a^2} \tag{A.8}$$

$$\int \frac{1}{x\sqrt{a+bx}}\,dx = \frac{1}{\sqrt{a}}\ln\left(\frac{\sqrt{a+bx}-\sqrt{a}}{\sqrt{a+bx}+\sqrt{a}}\right) \quad a>0 \tag{A.9}$$

$$\int \frac{1}{x^2 + a^2} dx = \frac{1}{a} \arctan\left(\frac{x}{a}\right) \qquad (A.10)$$

$$\int \frac{1}{(x^2 + a^2)^2} dx = \frac{1}{2a^3} \arctan\left(\frac{x}{a}\right) + \frac{x}{2a^2(x^2 + a^2)} \qquad (A.11)$$

$$\int \frac{1}{(x^2 \pm a^2)^{3/2}} dx = \frac{\pm x}{a^2\sqrt{x^2 \pm a^2}} \qquad (A.12)$$

$$\int \frac{x}{(x^2 \pm a^2)^{3/2}} dx = -\frac{1}{\sqrt{x^2 \pm a^2}} \qquad (A.13)$$

$$\int \frac{1}{(x^2 + b^2)\sqrt{x^2 + a^2}} dx = \qquad (A.14)$$

$$\frac{1}{b\sqrt{a^2 - b^2}} \arctan\left(\frac{x\sqrt{a^2 - b^2}}{b\sqrt{x^2 + a^2}}\right) \quad \text{for } b^2 < a^2$$

$$\int \sin(ax) dx = -\frac{1}{a} \cos(ax) \qquad (A.15)$$

$$\int \cos(ax) dx = \frac{1}{a} \sin(ax) \qquad (A.16)$$

$$\int \tan(ax) dx = -\frac{1}{a} \ln(\cos(ax)) \qquad (A.17)$$

$$\int \sin^2(ax) dx = \frac{x}{2} - \frac{\sin(2ax)}{4a} \qquad (A.18)$$

$$\int \cos^2(ax) dx = \frac{x}{2} + \frac{\sin(2ax)}{4a} \qquad (A.19)$$

$$\int \sin^3(ax) dx = -\frac{1}{3a}(\cos(ax))(\sin^2(ax) + 2) \qquad (A.20)$$

$$\int x \sin(ax) dx = \frac{1}{a^2} \sin(ax) - \frac{x}{a} \cos(ax) \qquad (A.21)$$

$$\int x^2 \sin(ax) dx = \frac{2x}{a^2} \sin(ax) - \frac{a^2 x^2 - 2}{a^3} \cos(ax) \qquad (A.22)$$

$$\int x \cos(ax) dx = \frac{1}{a^2} \cos(ax) + \frac{x}{a} \sin(ax) \qquad (A.23)$$

A. INTEGRAL FORMULAS

$$\int x^2 \cos(ax) dx = \frac{2x \cos(ax)}{a^2} \sin(ax) - \frac{a^2 x^2 - 2}{a^3} \sin(ax) \tag{A.24}$$

$$\int_0^\infty x^n e^{-ax} dx = \frac{n!}{a^{n+1}} \qquad a > 0, \quad n > 0 \tag{A.25}$$

$$\int_0^\infty e^{-ax^2} dx = \frac{1}{2}\sqrt{\frac{\pi}{a}} \qquad a > 0 \tag{A.26}$$

$$\int_0^\infty x^{2n} e^{-ax^2} dx = \frac{1 \cdot 3 \cdot 5 \cdots (2n-1)}{2^{n+1} a^n} \sqrt{\frac{\pi}{a}} \tag{A.27}$$

$$\int_0^\infty x^{2n+1} e^{-ax^2} dx = \frac{n!}{2a^{n+1}} \qquad a > 0, \quad n > -1 \tag{A.28}$$

$$\int_{-\infty}^\infty \delta(x) dx = 1 \tag{A.29}$$

$$\int_{-\infty}^\infty f(x) \delta(x) dx = f(0) \tag{A.30}$$

$$\int_{-\infty}^\infty f(x) \delta(x-a) dx = f(a) \tag{A.31}$$

B Trigonometric Identities

$$\sin^2 x + \cos^2 x = 1. \tag{B.1}$$

$$\sin(\alpha + \beta) = \sin\alpha\cos\beta + \cos\alpha\sin\beta \tag{B.2}$$

$$\cos(\alpha + \beta) = \cos\alpha\cos\beta - \sin\alpha\sin\beta \tag{B.3}$$

$$\sin(2\alpha) = 2\sin\alpha\cos\alpha \tag{B.4}$$

$$\sin\alpha + \sin\beta = 2\sin\frac{1}{2}(\alpha+\beta)\cos\frac{1}{2}(\alpha-\beta) \tag{B.5}$$

$$\cos\alpha + \cos\beta = 2\cos\frac{1}{2}(\alpha+\beta)\cos\frac{1}{2}(\alpha-\beta) \tag{B.6}$$

$$\sin^2\alpha = \frac{1}{2}(1 - \cos 2\alpha) \tag{B.7}$$

$$\cos^2\alpha = \frac{1}{2}(1 + \cos 2\alpha) \tag{B.8}$$

$$-\sin x = \cos\left(x + \frac{\pi}{2}\right) \tag{B.9}$$

$$\sin x = \cos\left(x - \frac{\pi}{2}\right) \tag{B.10}$$

$$e^{i\alpha} = \cos\alpha + i\sin\alpha \tag{B.11}$$

$$\sin\alpha = \frac{e^{i\alpha} - e^{-i\alpha}}{2i} \tag{B.12}$$

$$\cos\alpha = \frac{e^{i\alpha} + e^{-i\alpha}}{2} \tag{B.13}$$

C Useful Relations

Various Differentials:

System	Area	Volume
Polar	$r\,dr\,d\theta$	
Cylindrical	$R\,d\theta\,dz$	$r\,dr\,d\theta\,dz$
Spherical	$R^2 \sin\theta\,d\theta\,d\phi$	$r^2 \sin\theta\,dr\,d\theta\,d\phi$

Vector Identities:

$$\vec{A} \times (\vec{B} \times \vec{C}) = (\vec{A} \cdot \vec{C})\vec{B} - (\vec{A} \cdot \vec{B})\vec{C}$$

$$(\vec{A} \times \vec{B}) \times \vec{C} = (\vec{A} \cdot \vec{C})\vec{B} - (\vec{B} \cdot \vec{C})\vec{A}$$

$$(\vec{A} \times \vec{B}) \times (\vec{C} \times \vec{D}) = (\vec{A} \cdot (\vec{C} \times \vec{D}))\vec{B} - (\vec{B} \cdot (\vec{C} \times \vec{D}))\vec{A}$$

$$(\vec{A} \times \vec{B}) \cdot (\vec{C} \times \vec{D}) = (\vec{A} \cdot \vec{C})(\vec{B} \cdot \vec{D}) - (\vec{A} \cdot \vec{D})(\vec{B} \cdot \vec{C})$$

Series Expansions:

Binomial expansion:

$$(1 \pm x)^n = 1 \pm nx + \frac{n(n-1)x^2}{2!} \pm \cdots$$

Exponential expansion:

$$e^x = 1 + x + \frac{x^2}{2!} + \frac{x^3}{3!} + \cdots$$

D Relations Involving the Del Operator

$\vec{\nabla} = \frac{\partial}{\partial x}\hat{i} + \frac{\partial}{\partial y}\hat{j} + \frac{\partial}{\partial z}\hat{k}$

$\vec{\nabla} \cdot (\vec{\nabla} \times \vec{B}) = 0$

$\vec{\nabla} \times (\vec{\nabla} \times \vec{B}) = \vec{\nabla}(\vec{\nabla} \cdot \vec{B}) - \nabla^2 \vec{B}$

$\vec{\nabla} \cdot \phi\vec{A} = \phi\vec{\nabla} \cdot \vec{A} + \vec{A} \cdot \vec{\nabla}\phi$

$\vec{\nabla} \times \phi\vec{A} = \phi\vec{\nabla} \times \vec{A} + \vec{\nabla}\phi \times \vec{A}$

$\vec{\nabla} \cdot (\vec{A} \times \vec{B}) = \vec{B} \cdot (\vec{\nabla} \times \vec{A}) - \vec{A} \cdot (\vec{\nabla} \times \vec{B})$

$\vec{\nabla} \times (\vec{A} \times \vec{B}) = (\vec{B} \cdot \vec{\nabla})\vec{A} - (\vec{A} \cdot \vec{\nabla})\vec{B} + (\vec{\nabla} \cdot \vec{B})\vec{A} - (\vec{\nabla} \cdot \vec{A})\vec{B}$

$\vec{\nabla}(\vec{A} \cdot \vec{B}) = (\vec{A} \cdot \vec{\nabla})\vec{B} + (\vec{B} \cdot \vec{\nabla})\vec{A} + \vec{A} \times (\vec{\nabla} \times \vec{B}) + \vec{B} \times (\vec{\nabla} \times \vec{A})$

$\vec{\nabla} f(x) = \frac{df}{dx}\vec{\nabla} f$

E Del in Curvlinear Coordinates

Cylindrical:

$\vec{\nabla}\Psi = \frac{\partial \Psi}{\partial r}\,\hat{r} + \frac{1}{r}\frac{\partial \Psi}{\partial \phi}\,\hat{\phi} + \frac{\partial \Psi}{\partial z}\,\hat{z}$

$\vec{\nabla}\cdot\vec{A} = \frac{1}{\rho}\frac{\partial}{\partial \rho}(\rho A_\rho) + \frac{1}{\rho}\frac{\partial A_\phi}{\partial \phi} + \frac{\partial A_z}{\partial z}$

$\nabla^2\Psi = \frac{1}{r}\frac{\partial}{\partial r}\left(r\frac{\partial \Psi}{\partial r}\right) + \frac{1}{r^2}\frac{\partial^2 \Psi}{\partial \phi^2} + \frac{\partial^2 \Psi}{\partial z^2}$

$\vec{\nabla}\times\vec{A} = \frac{1}{r}\begin{vmatrix} \hat{r} & r\hat{\phi} & \hat{z} \\ \frac{\partial}{\partial r} & \frac{\partial}{\partial \phi} & \frac{\partial}{\partial z} \\ A_r & rA_\phi & A_z \end{vmatrix}$

$\hat{x} = \cos\phi\,\hat{r} - \sin\phi\,\hat{\phi}$

$\hat{y} = \sin\phi\,\hat{r} + \cos\phi\,\hat{\phi}$

$\hat{z} = \hat{z}$

Spherical:

$\vec{\nabla}\Psi = \frac{\partial \Psi}{\partial r}\,\hat{r} + \frac{1}{r}\frac{\partial \Psi}{\partial \theta}\,\hat{\theta} + \frac{1}{r\sin\theta}\frac{\partial \Psi}{\partial \phi}\,\hat{\phi}$

$\vec{\nabla}\cdot\vec{A} = \frac{1}{r^2\sin\theta}\left[\sin\theta\frac{\partial}{\partial r}(r^2 A_r) + r\frac{\partial}{\partial \theta}(\sin\theta A_\theta) + r\frac{\partial A_\phi}{\partial \phi}\right]$

$\nabla^2\Psi = \frac{1}{r^2\sin\theta}\left[\sin\theta\frac{\partial}{\partial r}\left(r^2\frac{\partial \Psi}{\partial r}\right) + \frac{\partial}{\partial \theta}\left(\sin\theta\frac{\partial \Psi}{\partial \theta}\right) + \frac{1}{\sin\theta}\frac{\partial^2 \Psi}{\partial \phi^2}\right]$

$\vec{\nabla}\times\vec{A} = \frac{1}{r^2\sin\theta}\begin{vmatrix} \hat{r} & r\hat{\theta} & r\sin\theta\hat{\phi} \\ \frac{\partial}{\partial r} & \frac{\partial}{\partial \theta} & \frac{\partial}{\partial \phi} \\ A_r & rA_\theta & r\sin\theta A_\phi \end{vmatrix}$

$\hat{x} = \sin\theta\cos\phi\,\hat{r} + \cos\theta\cos\phi\,\hat{\theta} - \sin\phi\,\hat{\phi}$

$\hat{y} = \sin\theta\sin\phi\,\hat{r} + \cos\theta\sin\phi\,\hat{\theta} + \cos\phi\,\hat{\phi} \quad \hat{z} = \cos\theta\,\hat{r} - \sin\theta\,\hat{\theta}$

F Physical Constants

Gravitational constant: $G = 6.67 \times 10^{-11}$ Nm^2/kg^2

Boltzmann's constant: $k_B = 1.38 \times 10^{-23}$ J/K

Gas constant: $R = 8.314$ J/(mol K)

Avogadro's number: $N_A = 6.02 \times 10^{23}$ particles/mol

Stefan–Boltzmann constant: $\sigma = 5.67 \times 10^{-8}$ W/(m^{2}K^4)

Mass of proton: $m_p = 1.67 \times 10^{-27}$ kg

Mass of electron: $m_e = 9.11 \times 10^{-31}$ kg

Fundamental unit of charge: $q = 1.60 \times 10^{-19}$ C

Permittivity of free space: $\epsilon_o = 8.85 \times 10^{-12}$ C^2/(Nm^2)

Permeability of free space: $\mu_o = 1.26 \times 10^{-6}$ Tm/A

Speed of light in vacuum: $c = 2.997925 \times 10^8$ m/s

Pressure at sea-level: $P = 1.013 \times 10^5$ Pa

G INDEX

Alinco alloys, 213
Amp, unit of, 144
American Wire Gauge Scale, 152
Ampère's Law, 186
Antennae, 274
Anticommutative property, 13
Atomic polarizability, 114

Basis vectors in cylindrical coordinates, 46
Biot-Savart law, 161
Bound charge, surface, 115
Bound charge, volume, 116
Brown and Sharpe scale, 152

Capacitance, 99
Capacitor, 100
Cartesian coordinates, 6
Charge, 55
Charge density, 64
Circumferential field, 27, 160
Clausius-Mossotti equation, 136
Coaxial cable, 193
Cofactor, 16
Cofactors, expansion by, 15
Compass, 143
Conductivity, 148
Conductivity, electrical, 146
Conductors, 56
Conservative field, 84
Continuity equation, 27
Coulomb, Charles, 56
Coulomb force, 57
Coulomb's law, 56
Curie temperature, 214
Current, 144
Current loop, 179

Curvilinear coordinates, 7
Cyclotron frequency, 167
Cylindrical coordinates, 43

Del operator, 24
Determinant of a matrix, 15
Diamagnetism, 198
Dielectric, 113
Dielectric breakdown, 124
Dielectric constant, 121
Dielectric strength, 124
Displacement current, 257
Displacement field, 126
Domains, ferromagnetic, 199
Domains, magnetic, 211
Dot product, 8
Drift speed, 148
Drift velocity, 145

Electrecs, 117
Electric dipole, 113
Electric dipole radiation, 274
Electric field, 62
Electric potential difference, 90
Electric susceptibility, 120
Electromagnets, 191
Electromagnetic induction, 221
Electromagnetic radiation, 261
Electrostatic fields, 77
Electron spin, 184
emf, 219
Energy density, of E & M wave, 264
Energy dissipation, conductors, 155
Energy in dielectrics, 134
Energy, in electric fields, 83, 105
Equivalent capacitance, 108
Equivalent inductance, 251

Faraday's law, 219
Ferromagnetism, 199
Flux, fluid flow, 26
Flux of a vector field, 34
Force, magnetic, 165
Free charge, 126
Fundamental Theorem for Divergences, 40
Fundamental theorem for curls, 40

Gauss's Law, 69, 70
Gauss's law, magnetic fields, 254
Gauss's theorem, 40
Gaussian pillbox, 74
Gradient, 24
Gradient, cylindrical coordinates, 48
Gradient, cylindrical coordinates, 50

H field, 203
Hard ferromagnetic, 213
Homogeneous equation, 32
Hydrogen atom, 183
Hysteresis, 211

Inductance, 229
Inductor, 229
Inner product for vectors, 8
Insulators, 56
Intensity, average of E & M wave, 265
Iron core transformer, 239
Irrotational vector field, 29

Joule heating law, 155

Laplace expansion, 15
Laplace's equation, 32, 93
Laplacian, 32
Lennard-Jones potential, 89
Lenz's law, 230
Line integrals, 37
Linear magnetic materials, 205
Lorentz force, 167

Magnet, 211
Magnetic circuit, 243
Magnetic dipole, 179
Magnetic poles, 180
Magnetic susceptibility, 206
Magnetism, 143
Magnetization, 210
Magnitude of a vector, 6
Mass spectrometer, 175
Matrix, square, 14
Maxwell's equations, 253, 254
Maxwell, James Clerk, 253
Minor, of a matrix, 16
Mobility, electron, 153
Molar magnetic susceptibility, 206
Motion in magnetic field, 166
Mutual inductance, 236, 238
Multi-load transformer, 242

Norm of a vector, 8
Nonhomogeneous equation, 32

Oersted, Hans Christian, 143
Ohm's law, 149
Ohms' law, for circuits, 154

G. INDEX

Orthogonal basis, 6
Orbital magnetic dipole, 183

Paramagnetism, 199
Permeability of free space, 58, 162
Permeance, 243
Permittivity, 121
Plane polarized wave, 260
Polarization, 114
Polarizability, electronic deformation, 114
Polarizability, ionic deformation, 115
Polarizability, orientational, 115
Polarization current, 267
Polarized atoms, 112
Poisson's equation, 32, 93
Poisson's equivalent distributions, 116
Potential energy, static charges, 85
Poynting vector, 265

Relative permittivity, 121
Reluctance, 243
Resistivity, 149
Retardation effects, 275
Right hand rule, magnetic fields, 160

Saturation, magnetic, 213
Scalars, 5
Scalar triple product, 19
Schwarz inequality, 8
Self inductance, 228
Sinks, 27
Skin depth, 273

Soft ferromagnetic, 213
Solenoid, 189
Solenoidal field, 27
Sources, 27
Spherical coordinates, 6
Static charges, 57
Static vector fields, 23
Stoke's theorem, 40
Sub-matrices, 16
Superposition, principle of, 58

Torque, on magnetic dipole, 182
Transformer equation, 239
Transformers, 239

Uniform circular motion, 166
Uniform vector field, 23
Unit vectors, 6

Vectors, 5
vector, cross product, 12, 14
Vector fields, 22
Vector valued function, 6
Velocity selector, 172
Voltage, 90

Wave equation, 259
Wave equation, damped, 271
Wire, 150
Work of Coulomb force, 83

Yukawa potential, 90

www.ingramcontent.com/pod-product-compliance
Lightning Source LLC
Chambersburg PA
CBHW072131170526
45158CB00004BA/1333